Humans and Automata

BEYOND HUMANISM: TRANS- AND POSTHUMANISM
JENSEITS DES HUMANISMUS: TRANS- UND POST-HUMANISMUS

Edited by / Herausgegeben von Stefan Lorenz Sorgner

Editorial Board:
H. James Birx
Irina Deretic
James J. Hughes
Andy Miah
Domna Pastourmatzi
Evi Sampanikou

VOL./BD. 7

Zu Qualitätssicherung und Peer Review der vorliegenden Publikation

Die Qualität der in dieser Reihe erscheinenden Arbeiten wird vor der Publikation durch den Herausgeber der Reihe geprüft.

Notes on the quality assurance and peer review of this publication

Prior to publication, the quality of the work published in this series is reviewed by the editor of the series.

Riccardo Campa

Humans and Automata

A Social Study of Robotics

Bibliographic Information published by the Deutsche Nationalbibliothek
The Deutsche Nationalbibliothek lists this publication in the Deutsche Nationalbibliografie; detailed bibliographic data is available in the internet at http://dnb.d-nb.de.

Cover illustration:
Rita de Muynck: Is the Future in the Past? Cyborg.
Audio-sculptural Installation.
Sheet, ideas, text, eartrumpets, spreaker, sound.
145 x 250 x 65 cm, 2011.
(© De Muynck,
Foto: Susanne Hesping)

Library of Congress Cataloging-in-Publication Data
Campa, Riccardo, author.
  Humans and automata : a social study of robotics / Riccardo Campa.
     pages cm – (Beyond humanism : trans-and posthumanism, ISSN 2191-0391 ; vol 7
= Jenseits des Humanisms : trans-und ; Bd 7)
  Includes bibliographical references and index.
  ISBN 978-3-631-66628-9 (print) – ISBN 978-3-653-05946-5 (e-Book) 1. Robotics–
Social aspects. 2. Human-machine systems. 3. Robots--Social aspects. 4. Robotics–
Moral and ethical aspects. I. Title.
  TJ211.49.C36 2016
  303.48'3–dc23
                             2015029765

ISSN 2191-0391
ISBN 978-3-631-66628-9 (Print)
E-ISBN 978-3-653-05946-5 (E-Book)
DOI 10.3726/978-3-653-05946-5

© Peter Lang GmbH
Internationaler Verlag der Wissenschaften
Frankfurt am Main 2015
All rights reserved.
Peter Lang Edition is an Imprint of Peter Lang GmbH.

Peter Lang – Frankfurt am Main · Bern · Bruxelles · New York ·
Oxford · Warszawa · Wien

All parts of this publication are protected by copyright. Any utilisation outside the strict limits of the copyright law, without the permission of the publisher, is forbidden and liable to prosecution. This applies in particular to reproductions, translations, microfilming, and storage and processing in electronic retrieval systems.

This publication has been peer reviewed.

www.peterlang.com

> At bottom, robotics is about us. It is the discipline of emulating our lives, of wondering how we work.
> Rod Grupen

Contents

Preface 11

Acknowledgments 21

1. Engineers and Automata 23
   1.1 A definition of 'robot' 23
   1.2 An historical overview 24
   1.3 The bottom-up approach to robotics 26
   1.4 The rise of social robots 29

2. Workers and Automata 37
   2.1 Artificial Intelligence and Industrial Automation 37
   2.2 Effects on the level of employment 40
   2.3 Social stratification and new generation robots 47
   2.4 The need for a new socio-industrial policy 49

3. Citizens and Automata 55
   3.1 Technology and unemployment 55
   3.2 Some methodological tools for scenario analysis 59
   3.3 The unplanned end of work scenario 62
   3.4 The planned end of robots scenario 64
   3.5 The unplanned end of robots scenario 67
   3.6 The planned end of work scenario 70
   3.7 An ethical judgement 74
   3.8 Conclusions 75

4. Roboethicists and Automata 77
   4.1 Roboethics: a discipline in statu nascendi 77
   4.2 A discipline concerned with futurabilia 80
   4.3 Roboethical codes 84
       4.3.1 Asimov's three laws of robotics 86
       4.3.2 The Euron Codex 89
   4.4 Evolution and legal responsibility 93
   4.5 Possible ethical problems in android robotics 98
       4.5.1 The NDR 114 Model 98
       4.5.2 The Galatea Model 101
       4.5.3 The Messalina Model 105
       4.5.4 The Gurdulù Model 106
       4.5.5 The Golem Model 107

5. Soldiers and Automata 109
   5.1 Defining robotic weapon 109
   5.2 Robots of the sky, the sea, and the land 110
       5.2.1 Sky Robots 113
       5.2.2 Sea Robots 116
       5.2.3 Land Robots 116
   5.3 The main functions of the military robots 121
   5.4 Main objections to the belligerent use of robots 125
       5.4.1 Noal Sharkey's plea 125
       5.4.2 Robotic wars as war crimes without criminals? 127
       5.4.3 Trivialization and multiplication of armed conflicts 127
   5.5 Analyses and propositions 134
       5.5.1 The Impracticability of the Moratorium 135
       5.5.2 Pragmatism as a remedy for undesired effects 137
       5.5.3 Voyeurism as an antidote to conflict escalation 139
       5.5.4 Correct information as a counterweight to alarmism ... 140

   5.6 Scenario analysis: dreams and nightmares 142
   5.7 Conclusions.. 146

Bibliography.. 151

Index of names.. 161

Preface

This book, as its title should quite clearly comunicate, is not on robotics as such, but rather on the social dimensions of robotics. So, let us start from a question: given that the author of this book is a sociologist, why is the research named 'a social study,' and not 'a sociological study,' of robotics? The main reason is that I explore different aspects of robotics, more specifically, the historical, economic, ethical, political, and futurological ones. The adjective that could possibly include all these aspects is 'social.' If sociology were the superscience originally intended by its founders in the 19th century (especially Comte and Spencer), we could still call this research a 'sociological study.' However, subsequently, sociology has strictly defined its object of study and research techniques, differentiating itself from other social sciences or disciplines such as history, economics, ethics, politics, and future studies.

The change in scope that characterizes the shift from classic to contemporary sociology has already been noticed by Florian Znaniecki (1934: v-iv) in his book *The Method of Sociology*. Znaniecki writes: "Now, sociology is passing through a crisis as deep as any science ever passed through. It was established as a synthetic science of 'society' or 'civilization,' using the results of several other sciences to draw such comprehensive generalizations as none of those sciences could or cared to draw for itself. It is changing into an analytic science investigating directly and independently particular empirical data, formulating its own results in a vast monographic literature, and not only avoiding hasty conclusions, but often mistrusting generalization more than other sciences do, and more than is good for any science."

Moreover, Max Weber notoriously proposed a rigid distinction between explanatory theories and normative theories, making of only the first the goal of sociology. This has been remarked, for instance, by the *Oxford Dictionary of Sociology*, which defines 'normative theories' as "hypotheses or other statements about what is right and wrong, desirable or undesirable, just or unjust in society," and points out that "the majority of sociologists consider it illegitimate to move from explanation to evaluation. In their view, sociology should strive to be value-free, objective, or at least to avoid making explicit value-judgements. This is because, according to the most popular philosophies of the social sciences, conflicts over values cannot be settled factually. Moral pronouncements cannot be objectively shown to be true or false, since value-judgements are subjective preferences, outside the realm of rational inquiry" (Marshall 2003).

Sometimes, this position has been taken too far, failing to recognize that Max Weber (2008, 46–7) was mainly talking about being neutral while teaching in the classroom – that is, avoiding behaving as a guru inside the walls of academy. To act as such a guru would have been 'unfair,' because students were asked to be passive listeners. Therefore, shifting from the objective ground of fact-statements to the subjective field of value-judgements could result in a kind of moral 'violence.' However, Weber openly stated that outside the classroom – i.e. in public conferences or in books – a social scientist could also engage in discussions about "what to do," and not only "what is or is not."

It is true that social science is different from social policy from a logical and methodological point of view, but it is also quite clear that the one needs the other. Applied sociology needs theoretical sociology, as robotic engineering needs theoretical physics. Contrarily to what most people assume, robotic engineering – as any type of engineering – is not value-free. It is value-laden no less than social policy or applied sociology, because it changes the world and the lifestyle of people, and very often it does it in a more radical way than social policies. One can argue, however, that social scientists tend to be more conscious than engineers about the political implications and the social consequences of their applications.

The main task of a sociologist is to reconstruct facts and unveil hidden mechanisms that establish a causal relation between certain actions and certain consequences. A sociologist does not venture into statements as to what humans ought to do on the basis of an ethical code or a political doctrine. If (s)he did, (s)he would be a moralist or a politician. Nevertheless – and this is what sociotechnology or social engineering consists of – a sociologist can still evaluate the lines of action from a chiefly technical point of view. For example, (s)he can say if the means M, adopted by agent A, in order to reach the objectives O1, O2, O3…On, is adequate or not in the light of the situation S in which the agent has to make a decision. In this case the evaluation is technical, not moral, because it is concerned with the means, not the ends. An Italian proverb speaks to this tense between the analyses of means and ends: "Non puoi avere la botte piena e la moglie ubriaca" ("one cannot have a full bottle and a drunk wife"). The sociologist will not tell a married man that he should give his wife a bottle of champagne, or that he would do better to save the money. But (s)he may be able to evaluate the efficiency of this man's strategies, whatever these may be, based on the observation that one cannot have a full bottle and a drunk wife. This is also what the physicist does when (s)he uses knowledge to modify reality, by elaborating theories that will be useful to build machines. In this case we call him an engineer, and certainly not a moralist, even if what he does has profound consequences on the life of a lot of people.

In spite of the fact that – given a certain definition of sociology – this work could be categorized as sociological, I decided to opt for a different title in order to avoid the endless epistemological and methodological discussions that surround sociology. We are not here to decide what sociology is or is not, what it should be or should not be, or if sociology is a real science or not. Labelling this work a "sociological study" could divert the attention from the real focus of this book: robots and their interactions with humans. Besides, I am comforted by the fact that interdisciplinary "social studies of science and technology" have already an established tradition inside the academic world.

This book has been composed by assembling and organizing essays and articles already published in English or Italian in scientific and academic journals. That is why the reader may find some concepts to be repeated in different parts of the book. This was a conscious decision made while compiling the various essays and articles, as each chapter is intended to be autonomous, with its own narrative structure. If concepts are repeated in different contexts that means that they are important within the given discussion and will helpful for the reader. This does not mean that those articles and fragments already published were left untouched. Robotics is rapidly evolving, therefore articles published two or three years ago often needed a 'face-lift,' that would take into account new direct observations or more recent secondary literature on the topic.

Chapter 1 is entitled "Engineers and Automata."[1] Its purpose is to examine some of the projects on which robotic engineers are presently working, and to explore the dreams and hopes connected with these undertakings. Here I offer some general information about robots and their origins, some technical information about the difference between the top-down and the bottom-up approaches to robotics, and I show how robotic engineers are incorporating a social understanding of robots. Finally, I explore the most recent literature on social robotics and argue that this interdiscipline is evolving in a direction that will soon require a systematic collaboration between engineers and social scientists.

Chapter 2 is entitled "Workers and Automata."[2] Its aim is to determine if there is a relation between automation and unemployment within the Italian socio-economic system. Technological unemployment is a long time debated problem.

---

1 Originally published in Italian as "Nascita e sviluppo della robotica" (Campa 2011, section 1.1.), with the exclusion of the section "The Rise of Social Robots," which is still unpublished.
2 Originally published in *Journal of Evolution and Technology* as "Workers and Automata. A Sociological Analysis of the Italian Case" (Campa 2014a).

One possible definition of technological unemployment is "unemployment due to our discovery of means of economising the use of labour outrunning the pace at which we can find new uses for labour" (Keynes 1930). The discussion about this phenomenon goes back at least to the beginning of the industrial revolution and the birth of Classical political economy.[3]

In Europe Italy has the second highest rate of robot density and the fourth highest rate in the world. Among the G7 it is the nation with the highest rate of youth unemployment. Establishing the ultimate causes of unemployment is a very difficult task, and the notion itself of 'technological unemployment' is controversial. Mainstream economics tends to relate the high rate of unemployment that characterizes Italian society with the low flexibility of the labour market and the high cost of manpower. Little attention is paid to the impact of artificial intelligence on rates of employment. With reference to statistical data, we will try to show that automation can be seen at least as a contributory cause of unemployment. In addition, we will argue that both Luddism and anti-Luddism are two faces of the same coin. In both cases attention is focused on technology itself (the means of production) instead of on the system (the mode of production). Banning robots or denying the problems of robotization are not effective solutions. A better approach would consist in combining growing automation with a more rational redistribution of income.

---

[3] One of the first scholars to study systematically technological unemployment was David Ricardo. Initially, Ricardo denies the problem. He maintains that the introduction of machinery is beneficial to all classes of society. He formulates what has been defined as "the first satisfactory statement of the theory of 'automatic compensation'" (Blaug 1958, 66). Other thinkers (among them: Say, Sismondi, and Malthus) oppose this view. Subsequently, "Ricardo retracted his former opinion on the subject" (Kurz 1984), shocking his own followers. Indeed, the idea "that the substitution of machinery for human labour, is often very injurious to the interests of the class of labourers" can be found in chapter XXXI, "On Machinery," in the third edition of Ricardo's *Principles*, published in 1821 (Ricardo 2004). This way, the idea of technological unemployment "marks its first appearance in respectable economic literature" (Kurz 1984; see also Lowe 1976, 250). As a matter of fact, the Luddites were denouncing this problem since long time, but only with Ricardo's classical economy, technological unemployment enters the 'world of ideas.' Karl Marx has also noticed that machinery did not free humans from labour, but rather caused unemployment on the one hand and the inhumane exploitation of those still employed on the other hand. Quite significantly, Marx praises Ricardo's "scientific impartiality and love of truth" (Marx 1976, 565). Similarly, Lowe (1954, 142) will characterize the chapter "On Machinery" by Ricardo as "a rare case of self-destructive intellectual honesty."

Chapter 3 is entitled "Citizens and Automata."[4] The aim of this chapter is to explore the possible futures generated by the development of artificial intelligence. Our focus will be on the social consequences of automation and robotization, with special attention being paid to the problem of unemployment. In spite of the fact that this investigation is mainly speculative in character, we will try to develop our analysis in a methodologically sound way. To start, we will make clear that the relation between technology and structural unemployment is still controversial. Therefore, the hypothetical character of this relation must be fully recognized.[5] Secondly, as proper scenario analysis requires, we will not limit ourselves to predict a unique future, but we will extrapolate from present data at least four different possible developments: 1) unplanned end of work scenario; 2) planned end of robots scenario; 3) unplanned end of robots scenario, and 4) planned end of work scenario. Finally, we will relate the possible developments not just to observed trends but also to social and industrial policies presently at work in our society, which may change the course of these trends.

Chapter 4 is entitled "Roboethicists and Automata."[6] Here we attempt to outline the core ideas of roboethics, that is, of ethical analysis as applied to robotics. First of all, we make some general remarks about the purpose and origins of this field, after which we move on to the normative aspects. We then define our ethical perspective as a whole and propose a 'rational' and 'pragmatic' approach to moral problems. Briefly, this approach consists in positing the 'good,' as understood as 'happiness,' as the fundamental value of ethics (in agreement with the Greek philosophical tradition), then rationally working out the behavioural norms that ought to lead to this 'good' and, finally, empirically and pragmatically assessing if respecting these norms *de facto* allows us to attain this end. If this is not the case it will be necessary to rationally reformulate the norms. We will then explore the main points of two well-known codes of roboethics: Asimov's "Three Laws of Robotics" and the "Euron Code." We have established that, given our rational and pragmatic standpoint, the Euron Code is the most suitable code for the regulation of military robotics. Essentially it boils down to five basic recommendations: *Safety* (control by humans); *Security* (preventing illegal use); *Privacy*

---

4   Originally published in *Journal of Evolution and Technology* as "Technological Growth and Unemployment: A Global Scenario Analysis" (Campa 2014b).
5   Indeed, a variety of factors have been indicated as determinants of unemployment (Gangl 2003, Malinvaud 1994, Reutter 2001, Rubart 2007, Sonnenberg 2014, Howell 2005, Stockhammer 2004, Gordon 2004, Groot 2004, Giugni 2009, Werding 2006, Salvadori & Balducci 2005, Ostrup 2003, Vroman & Brusentsev 2005, Hughes 2014).
6   Originally published in Italian as "Principi di roboetica" (Campa 2011, chapter 2).

(protection of data collected by the robot); *Traceability* (recording every activity carried out by the robot); and *Identifiability* (attribution of a secure and registered identity to every single robot). These recommendations are still compatible with a belligerent use of suitably regulated robots. On the contrary, Asimov's laws simply reject the possibility of using robots in warfare: indeed the first law states that "A robot may not injure a human being or, through inaction, allow a human being to come to harm" (without distinction between friends and enemies). Finally we tackle the legal mess arising from the problem of the juridical responsibility of robot behaviour, which can be fall under many subjects: the designer, the builder, the owner, the user, and – in an evolutionist perspective – the robot itself.

Chapter 5 is entitled "Soldiers and Automata."[7] The purpose of this chapter is to examine systems of military robots (or robotized weapons), with special emphasis on the ethical issues already raised by the design, construction and utilization of these technological objects, as well as on the possible future development of these systems of weaponry and of the new ethical problems they lead to. We will explore the field of military robotics in more detail, along the way pointing out the features of the models of robotic weapons currently in use on the battle field or in the stage of prototypes. We will give particular attention to the function of these machines and to the specific needs expressed by the soldiers using them. Our information comes from newspaper articles, books and documents compiled by military authorities. The *Unmanned System Roadmap* published by the United States Department of Defense stands out for its informative and futurological perspective, and is therefore particularly useful.

Then, we proceed by examining ethical problems of military roboethics. We go into further detail and apply the ethical norms that we worked out and discussed in chapter four to the robotized weapons presented and analysed in the first part of chapter five. Initially we reconstruct the main objections advanced by experts against a belligerent use of robots. First of all we outline Noal Sharkey's plea for an international moratorium, which has been echoed in the press. Then we take a look at the worries, expounded by some thinkers, that robots belong neither to the category of inanimate objects nor to that of sentient beings, and that there could therefore be war crimes without war criminals. Finally we analyse a number of current and hypothetical problems expounded by Peter W. Singer. Among these are a) the trivialization of war due to remotely piloted "Unmanned Aircraft Systems" (such as the "Predator"); b) the rise of "war porn" because robotic weapons

---

7   Originally published in Italian as "La robotica militare," "Problemi etici della robotica militare," and "Uno sguardo al futuro" (Campa 2011, section 1.2, chapters 3 and 4).

record war action, with the footage ending up online and thereby whetting sick voyeurism in the viewers; c) the possible multiplication of conflicts because of the indifference of ordinary people who will no longer have to give their lives, no longer be subjected to rationing, no longer pay war tax, no longer be called to vote for war intervention.

Once we have highlighted these various problems we come to the defence of our own standpoint. So, in the last pages of the book we will openly shift from an analytical-descriptive perspective to an axiological-normative one. Regarding the many proposals to ban or imposing moratoriums on military robots, among which Sharkey's is just the most well known, we observe that these suggestions are unrealistic and impracticable. Indeed the arms race is a typical case of the game theory known as "the prisoner's dilemma." Even if one could rationally argue in favour of the adoption by the various manufacturers of a very restrictive precautionary principle, the effective lack of mutual trust between nation states will induce them to go for the option that carries least individual (and not global) risk: namely, the construction of ever more powerful robotic weapons. In my humble opinion, it is less unrealistic – and certainly more advisable – to initiate negotiations that purport not so much to block the research and development of military robots, but to regulate their actual use on the battlefield. This sort of convention has already been agreed upon and respected in the case of other systems of weaponry. This suggests that this approach is more practicable. In addition it is certainly possible to lay down strict rules relative to the identification and the traceability of the robots that give exact information regarding to whom they belong, whose responsibility they are, where they are located, what they are doing and what they have done. This ought to limit the damage due to the possible illegal use of these weapons by organized criminals or terrorist groups.

As regards voyeurism and "war porn," we will see that new technologies amplify these problems, but do not cause them. Anthropology teaches us that man is no angel, but a morally ambivalent being: on the one hand he is capable of remorse, compassion, altruism, but on the other he still remains the most ruthless predator to have appeared on planet Earth. In particular, unlike many other predators, he has proven able to kill for sheer pleasure or in revenge. Nevertheless it is precisely man's ambivalent nature that should induce us to also consider the possibility that the footages of robot soldiers might act as an antidote to the multiplication of conflicts, sensitising public opinion, and causing improper use to stir indignation.

The fact that man is by definition no angel should make us view the question of "control" by humans in a different light. This first point of the Euron Code is indeed the one that seems to require the most critical thinking if it is to serve as a source of efficient implemental norms. The dogma of "human control" risks becoming an obstacle, unless one takes two fundamental questions into account: 1) it is necessary to clarify what one means by "human control" (remote control? A veto on certain actions? The ability to incapacitate the robot at any time? The ability to switch it off given to any human being or only to the legitimate owner, to the manufacturers, to the users, to agencies of law enforcement?); 2) it is always necessary to keep in mind that mobsters and terrorists are *humans* and may gain *control* over such technology.

To sum up, we propose a different approach, one that is statistical and pragmatic. Given that human beings make mistakes (via inexperience, friendly fire, cruelty, pointless violence, torture, etc.), and given that even in war "control by humans" might become impracticable (maximum robot autonomy might confer military superiority over the adversary), it is better to avoid fixating on this principle. I think that it is more rational to allow machines increasing room for autonomy and use statistics to evaluate whether they commit more or fewer mistakes than humans. This will allow for the development of more sophisticated prototypes and for the further development of functional autonomy.

Finally, we will venture into futurology, and try to imagine forms and functions of future robots, with a particular attention to possible military use. We will examine the reasons that make us believe where we are heading. Basic reasons are the optimal performance of current robots, the increasing faith of the army in these machines and the requirements of its commanders (that the Department of Defense promptly transmits to designers and manufacturers). The synergy between the army, politicians and the private manufacturing sector suggests that this trend is unlikely to change. It may, however, possibly slow down because of unforeseen developments (the financial crisis, the withdrawal from Iraq, etc.). The two main innovations of the future will be the miniaturization of 'unmanned systems,' linked to the development of nanotechnology, and the emergence of robots with ever more human-like shapes and behaviours, the so-called 'androids.' These technologies – anticipated by scientists, futurists and writers of science fiction – will bring about hitherto unseen opportunities and dangers.

We will also examine a series of reasons why the robotic arms race is unlikely to end in the extinction of the human race (as predicted by works of science fiction like *The Terminator* or *The Matrix* or in some recent newspaper articles). However, while the apocalyptic outcome may be improbable, it is certainly not impossible. Faced with this uncertainty, in order to ethically justify the choice of going ahead with the further development of this technology, we will rely on the logical insufficiency of the precautionary principle, which does not give enough credit to all the benefits that may come from the development of robotics. In addition, I will make a recommendation that can be summed up as follows: it is not advantageous to halt technological progress, including that of military robots, but it is rational to develop lines of parallel research, the goal of which is to design technologies that will be able to counterbalance those harmful outcomes. To put it as a slogan: *any drug can become a poison, therefore every poison needs to have its antidote.*

Acknowledgments

I am beholden to the Jagiellonian University in Krakow, Poland – my academic home – for having granted me the academic freedom required for intellectual exploration and heterodoxy. I am grateful to the Institute for Ethics and Emerging Technologies (IEET) for having started a stimulating debate on robotization and technological unemployment, and in particular to its director James Hughes. I thank the Military Centre for Strategic Studies (Centro Militare Studi Strategici – Ce.Mi.S.S.) of the Italian Ministry of Defence for having appointed me director of a research project on robotic weapons and hostile artificial intelligence. In particular, I thank Italian Air Force colonel Volfango Pierluigi Monaci for having read and reviewed my research. A debt of gratitude is also due to the Industrial Research Institute for Automation and Measurements (Przemy-słowy Instytut Automatyki i Pomiarów – PIAP) in Warsaw for having invited me to collaborate within their RoboScope seminar, which provided me the opportunity to learn much about the perspectives of engineers and military personnel on the use of robots within special operations.

I am grateful to Stefan Lorenz Sorgner for his sustained encouragement and his patience; Luciano Pellicani, Warren Breckman, Mario Bunge, Piotr Sztompka, Michael Burawoy, Neil Smelser, Jeffrey Alexander, Lech Witkowski, Kazimierz Krzysztofek, Krzysztof Frysztacki, Jarosław Górniak, Michel Kowalewicz, Maria Flis, and Anthony Giddens for stimulating discussions which have helped to broaden my knowledge on social and cultural processes.

I am particularly indebted to Catarina Lamm for having translated my Italian writings on robotics and for having encouraged me to publish a book in English on this topic. This book would not exist without her initiative. I am also grateful to Matt Hammond and Lucas Mazur for having proofread various parts of the book. It goes without saying that any remaining inaccuracies in the facts or in the style are my own.

Last, but not least, I thank my little son Leonardo for his spontaneous and contagious fascination with robots and machinery, and for his many questions about the future that he will inhabit.

I dedicate this book to the memory of my master and teacher Robert K. Merton (1910–2003) sociologist, methodologist, historian of ideas, and outstanding student of science, technology and society.

1. Engineers and Automata

1.1 A definition of 'robot'

The term 'robot' refers first of all to an automatic system capable of replacing people in the execution of complex tasks, founded on a sensory and kinetic interaction with the environment. The latter specification is required to distinguish it from a desktop. The definition must, however, be enriched with further detail because, as it stands, it risks including in the category of robot items such as vacuum cleaners and lawnmowers.

While 'automaton' and 'machine' are terms that have been in use for quite some time, the term 'robot' is relatively recent, and has its beginning not in the world of real technology, but in that of science fiction. Only subsequently did industrial workshops and science laboratories adopt the word and its object.

Indeed, the term was coined by the Czech writer Karel Čapek (2004). It appeared for the first time in a 1921 novel titled *R.U.R.* (*Rossum's Universal Robots*). If the term has a mysterious and fascinating ring to English ears, it should be stressed that in Slavic languages its etymological root simply refers to the concepts of work and of worker. In Czech 'robota' means 'labour,' with emphasis on the fact that, according to the context, it signifies non-voluntary, or even forced, labour. In Polish 'worker' translates as 'robotnik,' and in addition it suggests not so much the manager as the man working along the assembly line.[8]

Of course, the robot represents an evolution of the ancient automaton, designs of which can be found that go back to the time of Heron and the Museum of Alexandria (Heron 1976; Russo 2004, 95–141), and more generally of the machine in Antiquity, the Middle Ages and the Renaissance. Specific to the robot are its interaction with the environment, the fact that it is not rigidly predetermined, and its degree of autonomy. In this respect, it is different from certain industrial machines and from ordinary consumer electronics. The most common vacuum cleaners must be manually guided by humans, and do not interact autonomously with the environment. They are not even remote controlled. Dishwashers, laundry machines, and hair dryers operate on preset programs. They do not dynamically interact with their environment; they do not change their mode of action in response to changing circumstances.

---

8   'Pracownik' is the Polish word for any kind of worker.

We know that today one makes vacuum cleaners and lawnmowers with robotic features. Some latest generation vacuum cleaners navigate autonomously, stop when they run into a wall, and turn or reverse. Also, when low on battery, they are able to find a wall outlet[9] and recharge automatically. Such consumer electronics could be categorized as non-anthropomorphic robots.

Nevertheless we expect a lot more from robots, precisely because we are used to thinking of them as they are found in science fiction. What do we expect from them? If not exactly consciousness, then at least a great variety of behaviours. We expect them to be able to choose a line of conduct from among many possible, to be able to make a decision in order to solve a problem in a complex situation, to be able to "do the right thing." The smart vacuum cleaner sucks up dust and nothing else. But already our home PC is able to do many things, depending on the software we install. It is in principle universal. Not in the sense that it can do *anything*, but in the sense that there can be no a priori list of what *it will be able to do*.

As such the hardware, the processor, the machine is plastic, flexible and adaptable. Its behaviour changes with the software installed. In this, it is very much like our brain. Maybe this is also what we expect from robots properly speaking. We expect machines to be able to do different things, complex things, things that we could not have foreseen when we acquired them. This also goes for 'military robots.' Presumably, a commander would appreciate having a multifunctional machine at his command, rather than a mono-functional one, that is, a machine able to invigilate, patrol, de-mine fields, fight, help the wounded, transport materials, etc. The human soldier – if compared to the current hyper-specialized robot soldier – still has the advantage of some versatility. Nevertheless robots are extremely appreciated in warfare because they are able to do certain things that are too hard, dangerous, dirty and sometimes impossible for humans. In addition, everything leads one to think that future robotic development will take place along the lines of greater versatility, as happened with computers.

1.2 An historical overview

If the machine is merely required to interact dynamically with the environment, it is enough to endow it with sensors and servomechanisms, that is, with mechanisms of retroaction or feedback. On the contrary, if a wide range of behaviours is required, then it is necessary to fit it with ever more sophisticated computers

---

9   To be more precise, they use a self-charging home base which is plugged into the wall, and they find their home base by means of infrared sensors.

in terms of hardware and software. Had dynamic interaction been enough, we could have said that robots began with the 19th century Industrial Revolution. Indeed, the first systems of automatic control and regulation were found in the steam engine. However, we demand more than that and do not want to leave aside the ability to calculate, complex perception, movement in space and adaptability. Therefore, if we are to properly speak of robots (and not merely of automation), we have to wait until the arrival of electronics, cybernetics, and artificial intelligence. This is exactly the point where the robotic engineer Hans Moravec begins the history of his field.

Serious attempts to build thinking machines began after the Second World War. One line of research, called 'cybernetics,' used electronic circuitry imitating nervous systems to make machines that learned to recognize simple patterns, and turtle-like robots that found their way to recharging plugs. A different approach, named 'artificial intelligence,' harnessed the arithmetic power of post-war computers to abstract reasoning, and by the 1960s made computers prove theorems in logic and geometry, solve calculus problems and play good games of checkers. At the end of the 1960s, research groups at MIT and Stanford attached television cameras and robot arms to their computers, so 'thinking' programs could begin to collect information directly from the real world.

But things did not quite turn out as the American scientists had expected. "What a shock! While the pure reasoning programs did their jobs about as well and about as fast as college freshmen, the best robot control programs took hours to find and pick up a few blocks on a table, and often failed completely, giving a performance much worse than a six month old child" (Moravec 1995).

How can we explain the disparity between thinking programs and programs acting in the real world? For one thing, it is a disparity which persists today. Calculating power and computer performance increase exponentially, following Moore's Law, while progress in robotics, and more specifically in machine perception and mobility, is much slower. Computers defeat the most brilliant people on the planet in the game of chess and make calculations no human mind can do, while robot motion is still clumsy, much clumsier than that of toddlers and young animals. It would seem then that it is much easier to think than to perceive and act upon the world around. How come?

Moravec (1995) finds the explanation inside the laws of evolution: "For hundreds of millions of years, our ancestors survived by seeing and moving better than their competition, and became fantastically efficient. We rarely appreciate our monumental skill because it is commonplace, shared by every human being and most animals. On the other hand, rational thinking, as in chess, is a newly acquired skill, perhaps less than one hundred thousand years old. The parts of our brain devoted to it are not well organized, and we do it very poorly. We didn't realize how poorly until recently, because we had no competition to show us up."

We ourselves have created our competitors. Unbelievable as it may seem, we have managed to create beings whose rational thinking is better and faster than our own. We have not yet managed to create beings that can move or perceive better than we do. Therefore this is the most difficult task. Nature has had hundreds of millions of years to work on this problem, while we have only grappled with it for a little over a few decades. It is not hard to imagine that it is just a matter of time. Among other things, clumsy machine motion applies only to 'anthropomorphic' machines. The remote controlled drones that the army currently sends out to the battlefield do what humans can only dream of: fly.

And this is exactly the point. While Japanese engineers try to perfect the android or humanoid, that is the human-like robot, encountering all the problems that Moravec has pointed out, American, Israeli or Italian engineers pay more attention to matter than to form and mostly work on vehicles (like airplanes, cars, ships) whose performances have already been established, or they take their inspiration from animals whose locomotion is easier to imitate than that of bipeds. Moreover, if the ability to carry out a task on the battlefield is what matters, it is of little importance if our 'fellow soldier' has wheels instead of legs. However, if the problem of motion can be bypassed, there remains the problem of perception, decision, of action broadly speaking.

1.3 The bottom-up approach to robotics

According to Moravec, the best approach to solve these technical problems is not the top-down approach, that is, from the mind to the body, of artificial intelligence, but the bottom-up approach that follows the steps of biological evolution, and that Moravec argues is the 'robotic' approach strictly speaking. Hence it is preferable to first study these living creatures and their evolution, and then reconstruct them bit by bit with non-biological devices.

In order to give sight to robots it is not enough to stick a video camera to a computer. It is necessary to understand and reconstruct human and animal vision, the functioning of retinal nerve cells and of the neurones that reinterpret the message. Incidentally, this is the mechanism of our own nervous system that we best understand. Moravec (1995) explains: "By comparing the edge and motion detecting circuitry in the four layers of nerve cells in the retina, the best understood major circuit in the human nervous system, with similar processes developed for computer vision systems that let robots see, I've estimated that it would take a billion computations per second, like an average supercomputer, to do the job of the human retina. By extrapolation, it will take the power of ten thousand supercomputers or a million personal computers to emulate a whole brain."

It should be added that the supercomputer is not the one standing on our desk. That is why, until now, the path of robotics has been long and winding. However, this need not be a cause for pessimism because progress in the different sciences relevant to robotics (computer science, electronics, biology, mechanics) is truly impressive. The systematic convergence of these areas of knowledge promises to give us robots, and perhaps also agile and intelligent androids, within a time frame that is not necessarily biblical.

Moravec at least is optimistic. This also because he is convinced that he has finally understood what the right approach is. In less than fifty years, according to him, we shall have computers able to compete, in complexity and calculating power, with the human brain, not just at the level of the neo-cortex's rational performance, but also at that of sensory and motor performances. He wrote in 1993: "The best robots today are controlled by computers just powerful enough to simulate the nervous system of an insect, cost as much as houses, and so find only a few profitable niches in society (among them, spray painting and spot welding cars and assembling electronics). But those few applications are encouraging research that is slowly providing a base for a huge future growth. Robot evolution in the direction of full intelligence will greatly accelerate, I believe, in about a decade when the mass-produced general purpose, *universal* robot becomes possible. These machines will do in the physical world what personal computers do in the world of data—act on our behalf as literal-minded slaves" (Moravec 1993).

This decade has come and gone and the mass-production of the universal robot is still to be seen. The media however does show us really extraordinary humanoid prototypes, produced by Japanese electromechanical companies – the most famous being probably Honda's Asimo. It still does not move like a gazelle, and it neither thinks nor speaks like a human, but little by little it appears to be heading that way. If it seems risky to make predictions, it seems just as obvious that the goal is indeed not beyond human possibilities.

Another step that we are taking only now is introducing the ability of robot learning. The *Babybot,* realized by a group of academics and researchers of the Faculty of Engineering of Genoa University, is exemplary here and has been much spoken about in the Italian and foreign press (Lira 2015). More precisely, the team of *Lira Lab*[10] has given life to a robot 'cub' in order to improve our understanding

---

10 Giulio Sandini (director), Giorgio Metta, Riccardo Manzotti, Vincenzo Tagliasco, Lorenzo Natale, Sajit Rao, and Carlos Beltran.

of the mechanisms of the human brain. It is not just that the robot does many things because its 'brain' can contain many programmes, but it can act in various ways because it has learned these behaviours through interacting with humans – that is because the robot is fitted with programs that allow it to register what it sees, hears, touches, and these memories then change into a stimulus to action. Action which thus becomes unpredictable.

When the journalist Antonella Polidori (2004) asked the researchers if Babybot has ever surprised them by doing something it had not been precisely programmed to do, Professor Giulio Sandini's reply was: "It happens very often; but it is part of how we work. We don't program the robot to do something in only one way. We give it an input to which the machine responds autonomously, and it is able to self-correct. If, hypothetically, it falls when grasping an object, Babybot learns from its mistake and tries again until it succeeds to grasp it correctly. Here is another example: although it is programmed to pay attention only to coloured objects, if it is stimulated long enough with grey things, it learns to respond to those as well."

As regards military robots, one of the problems most often encountered on the battlefield is the robot's inability to understand what the obstacle in front consists of. It perceives a barrier, but it doesn't know if it is a wall (meaning the kind of thing it must circumvent) or a hedge (meaning the kind of thing it can just go through unharmed). Therefore it performs a lot of unnecessary tasks in order to complete the mission. A robot capable of learning from its mistakes, exactly like a child, could improve its fighting skills with experience. Hence, a commander might prefer a 'veteran' robot, one that has already completed some military campaigns, to a 'newbie' just out of the assembly line, precisely as happens with human soldiers.

The learning mechanisms of these models is therefore very similar to that of animals and humans. This stage has been reached by some prototypes, among which are found biomechanical hybrids, meaning that they also contain neurones of a biological origin. The most known example is probably the Animat (Potter, DeMarse, Wagenaar, and Blau 2001). The next step, as foreseen by many writers of science fiction and by some currently working robotic engineers, first among them Moravec, is the appearance of a real and true artificial mind. Accepting a rigorously materialistic ontology, and taking it to its extreme consequences, many researchers are inclined to think that the mind is ultimately understandable to science and can therefore be emulated by technology.

Sandini however insists that the road ahead is still a long one: "We are still a very long way from 'building' a child! It is indeed while we try to build a system that can develop autonomously that we realize how far we still are from understanding some of the processes that transforms the defenceless infant into an independent adult." But, as history tells us from Heron to Alan Turing, passing through Leonardo Da Vinci and Guglielmo Marconi, it certainly takes more than obstacles to discourage an engineer. "My engineering mind – says Sandini – incites me to ponder not so much over obstacles as over how to overcome them" (Polidori 2004).

To conclude, for the time being our discussion about a future conscious or semi-conscious robot is purely academic, but we think that the discussion about the next possible step is as important as it is necessary. In other words, our priority is to study the robots already in existence and the social problems that they already raise. However, we cannot completely avoid futurological speculations, given the 'evolutionary' character of robots and the associated problems.

1.4 The rise of social robots

Robots are evolving. This is almost a truism. Therefore, in order to be truly meaningful, this observation needs to be more specific. The social consequences of robotics will depend to a significant degree on *how* and *in which direction* this discipline is evolving. That is why it could be instructive – for both sociologists interested in the social aspects of robotics and engineers interested in cooperating with sociologists – to have a closer look at some of the technical aspects of new generation robots.

One of the main sources of information about robotic trends is a book series published by Springer and edited by Bruno Siciliano and Oussama Khatib. As Siciliano (2013, v) states, "robotics is undergoing a major transformation in scope and dimension. From a largely dominant industrial focus, robotics is rapidly expanding into human environments and vigorously engaged in its new challenges. Interacting with, assisting, serving, and exploring with humans, the emerging robots will increasingly touch people and their lives." Siciliano noticed that the most striking advances happen at the intersection of disciplines. The progress of robotics does not have an impact only on the robots themselves, but also on other scientific disciplines. At the same time, these different research areas are sources of stimulation and insight for the field of robotics. Biomechanics, haptics, neurosciences, virtual simulation, animation, surgery, and sensor networks are just a few examples of the kinds of disciplines that stimulate and benefit from robotics research. Let us now explore a few examples in greater detail.

In 2013, four engineers – Jaydev P. Desai, Gregory Dudek, Oussama Khatib, and Vijay Kumar – edited a book entitled *Experimental Robotics*, a collection of essays compiled from the proceedings of the 13th International Symposium on Experimental Robotics. The main focus of many of these pieces is the problem of interaction and cooperation between humans and robots, and it is frequently argued that the effectiveness and safety of that cooperation may depend on technical solutions such as the use of pneumatic artificial muscles (Daerden and Lefeber 2000). Moreover, each technical device has advantages and disadvantages. For example, one may gain in effectiveness but lose in safety, or vice versa. In the article "Motor vs. Brake: Comparative Studies on Performance and Safety in Hybrid Actuations," evidence is presented that shows how the hybrid actuation with pneumatic artificial muscles and a brake "can be a competitive solution for the applications that require high efficiency, but accept a relatively low control performance, for example, a waist joint" (Shin, Yeh, Narita, and Khatib 2013, 101–102). The authors also specify that "[h]uman-centered robotics draws growing interest in inherently safe actuations for robots to cooperate with humans. Notable achievements are [the] series elastic actuator, variable impedance actuator, and distributed macro-mini actuation. In addition, considerable research has employed pneumatic artificial muscles (PAMs) for their high force-to-weight ratios and inherent compliances. Low output impedance of PAMs over a wide frequency range enables PAMs to reduce large impact forces during unforeseen collisions." On the other hand, "PAM's limited control performance prevents it from being more widely used. Due to their air compressibility and viscous/coulomb friction in a braided shell, force and position control bandwidths are limited."

It is important to keep these aspects in mind, as it is often the case that both technophiles and technophobes tend to anticipate fantastic or catastrophic developments, without considering the incremental, long and painstaking work on robotics which lay behind and ahead. There are many small problems like those surrounding PAMs mentioned above that need to be solved before we may start seeing NDR-114 from the film *Bicentennial Man* or Terminator-like machines walking around on the streets.

Another direction in which robotics is moving is that of small and even smaller automatic machines, such as: millirobots, microrobots, and nanorobots. In the Siciliano and Khatib series, there is an interesting book entitled *Small-Scale Robotics. From Nano-to-Millimeter-Sized Robotic Systems and Applications*, edited by Igor Paprotny and Sarah Bergbreiter (2014)[11]. In the "Preface," the editors make explicit the impact that science fiction has had on this area of research: "In the

---

11 The book contains selected papers based on presentations from the workshop "The Different Sizes of Small-Scale Robotics: from Nano-, to Millimeter-Sized Robotic Systems and Applications," which was held in conjunction with the International Conference on Robotics and Automation (ICRA 2013), in May 2013 in Karlsruhe, Germany.

1968 movie *The Fantastic Voyage*, a team of scientists is reduced in size to microscale dimensions and embarks on an amazing journey through the human body, along the way interacting with human microbiology in an attempt to remove an otherwise inoperable tumor. Today, a continuously growing group of robotic researchers [is] attempting to build tiny robotic systems that perhaps one day can make the vision of such direct interaction with human microbiology a reality."

Smaller-than-conventional robotic systems are described by the term 'small-scale robots.' These robots range from several millimeters to several nanometers in size. Applications for such robots are numerous. They can be employed in areas such as manufacturing, medicine, or search and rescue. Nonetheless, the step from imagination to realization, or from science fiction to science, is not a small one. There remain many challenges that need to be overcome, such as those related to the fabrication of such robots, to their control, and to the issue of power delivery.

Engineers regularly compare the capabilities of robotic systems, including small-scale robots, to those of biological systems of comparable size, and they often find inspiration in biology when attempting to solve technical problems in such areas as navigation and interactive behavior (Floreano and Mattiussi 2008, 399–514; Liu and Sun 2012; Wang, Tan, and Chew 2006). "The goal of small-scale robotics research is often to match, and ultimately surpass, the capabilities of a biological system of the same size. Autonomous biological systems at the millimeter scale (such as ants and fruit flies) are capable of sensing, control and motion that allows them to fully traverse highly unstructured environments and complete complex tasks such as foraging, mapping, or assembly. Although millimeter scale robotic systems still lack the complexity of their biological counterparts, advances in fabrication and integration technologies are progressively bringing their capabilities closer to that of biological systems" (Paprotny and Bergbreiter 2014, 9–10).

Presently, the capabilities of microrobotic systems are still far from those of microscale biological systems. Indeed, "biological systems continue to exhibit highly autonomous behavior down to the size for a few hundred micrometers. For example, the 400μm dust mite is capable of autonomously navigating in search for food and traversing highly unstructured environments. Similar capabilities can be found in *Amobeaproteous* or *Dicopomorpha zebra*" (Ibid., 9–10). On the contrary, microrobotic systems have only limited autonomy; they lack independent control as well as on-board power generation. In spite of the stark performance differences between biological systems and small-scale robots, engineers are far from being resigned to second place. Rather, they think that "these gaps highlight important areas of research while demonstrating the level of autonomy that should be attainable by future robotic systems at all scales" (Ibid., 10–11). Such statements speak to the optimistic mindset of engineers.

Another technical problem that also presents obvious ethical and social consequences is that of safety. An inspiring a book on the issue of safety in robotics is Sami Haddadin's *Towards Safe Robots: Approaching Asimov's 1ˢᵗ Law* (2014). Haddadin points out that the topic of research called Human-Robot Interaction is commonly divided into two major branches: 1) cognitive and social Human-Robot Interaction (cHRI); 2) physical Human-Robot Interaction (pHRI). Here is how he defines the two fields. cHRI "combines such diverse disciplines as psychology, cognitive science, human-computer interfaces, human factors, and artificial intelligence with robotics. Cognitive Human-Robot Interaction intends to understand the social and psychological aspects of possible interaction between humans and robots and seeks" to uncover its fundamental aspects. On the other hand, pHRI "deals to a large extent with the physical problems of interaction, especially from the view of robot design and control. It focuses on the realization of so called human-friendly robots by combining in a bottom-up approach suitable actuation technologies with advanced control algorithms, reactive motion generators, and path planning algorithms for achieving safe, intuitive, and high performance physical interaction schemes" (Haddadin 2014, 7).

Safety is obviously not a novel problem in robotics, nor in engineering more generally. It has been a primary concern in pHRI, since in this field continuous physical interaction is desired and it continues to grow in importance. In the past, engineers mainly anticipated the development of heavy machinery, with relatively little physical Human-Robot Interaction. The few small robots that were able to move autonomously in the environment and to interact with humans were too slow, predictable, and immature to pose any threat to humans. Consequently, the solution was quite easy: segregation. Safety standards were commonly tailored so as to separate the human workspace from that of robots.

Now the situation has changed. As Haddadin (2014, 7) put it, "due to several breakthroughs in robot design and control, first efforts were undertaken recently to shift focus in industrial environments and consider the close cooperation between human and robot. This necessitates fundamentally different approaches and forces the standardization bodies to specify new standards suitable for regulating Human-Robot Interaction (HRI)." These breakthroughs, and in particular the developments of cHRI, have opened the road to a new subdiscipline, or – if one prefers – a new interdisciplinary field: Social Robotics. In spite of the fact that the name appears to speak to a hybrid between the social sciences and engineering, at present, this subdiscipline is mainly being cultivated by engineers, although with a 'humanistic' sensitiveness.

In their book entitled *Human-Robot Interaction in Social Robotics* (2013), Takayuki Kanda and Hiroshi Ishiguru explain quite well the nature of the paradigm change the has accompanied the shift from industrial robots to interactive robots. They remind us that, up to recent times, robotics has been characterized by two main streams of research: navigation and manipulation. The first is the main function of autonomous mobile robots. The robot "observes the environment with cameras and laser scanners and builds the environmental model. With the acquired environmental model, it makes plans to move from the starting point to the destination" (Ibid., 1). The other stream in early robotics has been manipulation, as exemplified by research on robot arms. Like a human arm, the robot arm is often complex and therefore requires sophisticated planning algorithms. There are countless industry-related applications for both navigation and manipulation, and over the last several decades innovations in these research areas have revolutionized the field. Two different academic disciplines have been competing to solve the problems related to navigation and manipulation: Artificial Intelligence and robotics *strictu sensu*.

According to Kanda and Ishiguru (2013, 1–2), robotics now needs to engage in a new research issue: interaction. "Industrial robotics developed key components for building more human-like robots, such as sensors and motors. From 1990 to 2000, Japanese companies developed various animal-like and human-like robots. Sony developed AIBO, which is a dog-like robot and QRIO, which is a small human-like robot. Mitsubishi Heavy Industries, LTD developed Wakamaru. Honda developed a child-like robot called ASIMO. Unfortunately, Sony and Mitsubishi Heavy Industries, LTD have stopped the projects but Honda is still continuing. The purpose of these companies was to develop interactive robots (…)."

Social robotics is gaining in importance because mobile robots are increasingly required to perform tasks that necessitate their interaction with humans. What is more, such human-robot interactions are becoming a day-to-day occurrence. Japanese companies tend to develop humanoids and androids due to their strong conviction that machines with a human-like appearance can replicate the most natural of communicative partners for humans, namely other humans. In the words of Kanda and Ishiguru (2013, 5), the strongest reason for this research program is "in the human innate ability to recognize humans and prefer human interaction. The human brain does not react emotionally to artificial objects, such as computers and mobile phones. However, it has many associations with the human face and can react positively to resemblances to the human likeness."

Appearance is just one of the problems related to the social acceptance of robots. Verbal interaction is equally important. A group of researchers have recently edited a book entitled *Social Robotics* that presents interesting developments in the direction of improved HRI (Mutlu, Bartneck, Ham, Evers, and Kanda 2011). This volume collects the proceedings of the third International Conference on Social Robotics (ICSR), located in Amsterdam, The Netherlands, November 24–25, 2011.[12] In one of the contributions, Złotowski, Weiss, and Tscheligi (2011, 1–2) clearly explain the nature of this general field of research, as well the methodology that tends to be used. To begin, they underline that "the rapid development of robotic systems, which we can observe in recent years, allowed researchers to investigate HRI in places other than the prevailing factory settings. Robots have been employed in shopping malls, train stations, schools, streets and museums. In addition to entering new human environments, the design of HRI recently started shifting more and more from being solely technologically-driven towards a user-centered approach." Indeed, these particular researchers are working on a project called Interactive Urban Robot (IURO), "which develops a robot that is capable of navigating in densely populated human environments using only information obtained" from pedestrians it encounters.

Two key concepts in such research are 'scenario' and 'persona.' These design tools are already quite popular in Human-Computer Interaction (HCI). Now the approach based on these tools has been exported and adopted in HRI. "Scenarios are narrative stories consisting of one or more actors with goals and various objects they use in order to achieve these goals. Usually the actors used in scenarios are called personas. The main goal of personas is to ensure that the product being developed is designed for concrete users rather than an abstract, non existing 'average user.' Often, more than one persona is created in order to address the whole spectrum of the target group" (Ibid.).

The interesting aspect of Social Robotics is that researchers – even when they are basically trained as engineers – have to adopt a sociological or psychological perspective in order to create personas. This happens because the process of persona creation starts with the identification of key demographic aspects of the human populations of interest. In the case mentioned earlier, this particularly concerns pedestrians. In work on robot-pedestrian interaction, the researchers analyzed the profession, education, age range, and language skills of selected pedestrians and then augmented this information with pedestrian interviews; "this information was then enriched by the data obtained during interviews where we asked participants why they approached specific pedestrians. Not surprisingly,

---

12   Equally interesting are the volumes related to the previous and the following conferences. See: Ge Shuzhi, Li, Cabibihan, and Tan 2010; Ge Shuzhi, Khatib, Cabibihan, Simmons, and Williams 2012; Herrmann, Pearson, Lenz, Bremner, Spiers, and Leonards 2013.

we found that one of the most important factors, which impacts the successfulness of the interaction, was whether the encountered person was a local or not" (Ibid.).

It is not difficult to predict that as robots become more and more sophisticated, engineers will need the systematic help of trained sociologists and psychologists in order to 'create' personas and scenarios and to 'teach' humanoids how to behave in various circumstances. In other words, the increased interaction between mobile robots and humans is paving the way for increased interaction between *social robotics* – the study of HRI undertaken by engineers – and *robot sociology* – the study of the social aspects of robotics undertaken by social scientists.

2. Workers and Automata

2.1 Artificial Intelligence and Industrial Automation

The concept of 'artificial intelligence' is a vast one which includes all the forms of thinking produced by artificial machines. The concept of AI is therefore strongly related to that of automation, that is, of machines behaving autonomously, albeit in response to certain inputs and in the presence of programs. Any inorganic machine conceived and construed by humans – be they desktop computers or semi-mobile robots, dishwashers or power looms – and able to carry out the tasks that humans carry out using their own intelligence is an automaton. In other words, "a certain category of sets of elements are 'universal' in the sense that one can assemble such elements into machines with which one can realize functions which are arbitrary to within certain reasonable restrictions" (Minsky 1956). Given this definition, it follows that all functioning automata are endowed with a certain degree of artificial intelligence. The refrigerator is less intelligent than a PC, in more or less the same way as an insect is less intelligent than a vertebrate. And some do not hesitate to compare the various forms of organic and inorganic intelligence (Moravec 1997).

Automation is therefore not something new that has arisen in the last few years, but the fruit of a long and slow historical process that can be taken back to the mechanical calculators of Charles Babbage or Blaise Pascal, if not all the way to Heron's automata. Therefore whoever has a more revolutionary conception of artificial intelligence feels the need to introduce a distinction between weak AI and strong AI – a distinction that has a philosophical dimension and touches on matters such as the functioning of the brain and the ontology of the mind. This however will not be the topic of this chapter. Rather our intention here is to tackle the sociological aspects of artificial intelligence – whether it is understood as weak or strong, discrete or gradualistic. In other words, we intend to analyse the social, political, and economical consequences of the production and use of automata or thinking machines. Let us just set some temporal and spatial limits to our analysis. We will be looking at the artificial intelligence of the third industrial revolution (Campa 2007), which can be situated in the last decades of the 20th century and in the first decade of the 21st. In this period, automation is identified in particular with computerization and robotization. And we will chiefly be looking at Italy, which can in any case be viewed as exemplary, given that it is still among the first seven industrial powers of the planet and is one of the leading countries in the world for 'robot density.'

One of the most systematic applications of electronic calculators and of robots has so far been found in industrial plants. Microprocessors are omnipresent. Personal computers are found in every home and in every office. There is no institution that does not entrust some of its task to AI in some form or other. However, it is in the manufacturing industry that one observes some macroscopic social effects of the emergence of this technology.

We have all seen at least once robots that paint, weld and assemble cars, as well as electronic products such as radios, TV sets, and computers. These are the so-called industrial robots, which in advanced technological societies have come to work alongside and, in many cases, replace the worker on the assembly line. The first industrial robots appeared in the fifties, but it is only in the seventies that their presence in Italian plants began to become significant. They were steel constructions of impressive dimensions, endowed with a rudimentary electronic brain, faculties of perception, servomechanisms and hydraulic engines. The first generation industrial robots were slow and not particularly intelligent, and therefore their work was limited to tasks that do not require a high precision, like paint spraying and car body welding. Precision work was still done by humans. However, as could be foreseen, the situation changed quickly and in the eighties one could see robots able to assemble complex electronic circuits, inserting and welding the devices in a matter of seconds and without errors.

Industrial robots become ever more anthropomorphic. Their degree of freedom[13] increases their precision, velocity and load capacity. In the car and heavy industry they have little by little taken over other tasks that require precision such as piercing, grinding, milling, cutting, but also palletization and stockpiling. Nowadays they are endowed with laser devices and visual systems that allow them to operate with millimetric precision.

If it is the United States, as producers of the largest number of robots, that show the way, Japan also makes a massive entry into the sector from the seventies and onwards. Indeed, among the characteristic aspects of the third industrial revolution there is also the reorganization of the processes of production, with the computerization and automation of the entire factory, with Toyota as the true pioneer. It is not by chance that one tends to oppose Toyota's model to the model of organization based on the assembly line developed by Ford and Taylor. According

---

13 By 'degree of freedom' of an industrial robot, one means the number of axes of movement (in other words, the quantity of particular movements) that the machine is able to perform. The degree of freedom goes from 3-4 for the simplest robots to 9-10 degrees in the case of more complex ones. For comparison it is considered that the human hand has 23 degrees of freedom.

to Cristiano Martorella (2002) "thus the Japanese industrial revolution has transformed the factory into an information system and has freed man from mechanical work, transforming him into a supervisor of productive processes. This takes place in a period in history that sees the transition from the industrial society to the post-industrial society. This epochal turning point will be well understood once the transition to the society of services and information will be complete."

Italy contributes as well. FIAT is the first Italian company to make massive use of industrial robots. In general this country tends to import digital electronics from abroad, having lost foothold in this field, particularly after the crash of Olivetti in 1997. However, in robotics one sees very interesting exceptions to this rule. For example *Robogate* is an Italian invention that has since been adopted by the entire car industry.[14]

We will not enter into technical detail that the reader may find in handbooks (Kurfess 2012, Siciliano and Khatib 2008). Rather let us cast a rapid glance at the magnitude of the process of robotization in the industry. As the Italian newspaper *La Repubblica* stresses, "first generation robots, those that work in the industry all over the planet, number over a million: 350,000 in Japan alone, 326,000 in Europe. In Italy, for every 10,000 persons employed in the industry, there are over 100 robots, a number that makes our nation one of the first in the world in this sector. They are used above all for mechanical work, in welding and in working with plastics. And their price continues to fall: a robot purchased in 2007

---

14 Fiat has installed its Robogate equipment in 1978. As *New Scientist* (1980, 247) reported, "each system comprises a series of robot 'cells,' each of these containing two to four robots, which are arranged at intervals several metres along the line. The four basic parts to be welded are loaded onto a transporter, a low platform the size of a large tabletop. The transporter glides between the robot cells. Its motor is activated by signals passed along wires underneath the floor and detected by electromagnetic induction sensors. Movement of the transporter is controlled by a central computer." Flexibility is the main feature of the system: "engineers can change the system's parameters, while it is working on one model, to make a new design of car. To do this, engineers alter the software in both the robots and the central computer that controls the whole system; and make some changes to hardware, such as installing new gates for different car bodies. In less advanced robot installations, like those at Longbridge and and in many car factories in the US, operators do not have the benefit of this flexibility." As a consequence, the Fiat "plant requires only two men to run it, compared with 100 in a plant in which the welding is done by hand." *New Scientist* also reported that "Fiat has already sold one Robogate system to Chrysler in the US… General Motors is also interested in buying the system."

may cost a quarter of the price of the same robot sold in 1990. And if its yearly cost was 100 in 1990, today it is not above 25" (Bignami 2007).

More precisely, Italy is the second country in Europe and the fourth in the world regarding robot density, as shown by a more accurate study by the UNECE (2004, 2005). There are already more than 50000 units and the number continues to increase.

2.2 Effects on the level of employment

Thanks to censuses from the ISTAT, we are able to make very accurate comparisons between the growth of automation on the one hand, and the effects on employment on the other. If one leaves aside the census of the factories of the Kingdom of Italy that goes back to 1911, ISTAT has done nine censuses relative to industry, commerce and services (1927, 1937–39, 1951, 1961, 1971, 1981, 1991, 2001, 2011).

The first data of the 9th census (2011) were presented on 11th July 2013. It is not always easy to directly compare the statistics because with time the techniques of survey and the categories under scrutiny have changed: until 1971 the focus was 'industry and commerce,' while from 1981 it is on 'industry and services.' Statistical series have been 'harmonized' however, enabling an overall reading of the data. In addition, we are interested now in the manufacturing industry and consequently shifting the focus from commerce to services is of marginal relevance. Let us begin nonetheless with a comparison between the last three complete series of statistics (1981, 1991, 2001) that are fairly homogenous. The table concerns the absolute data:

Table 1 – Companies and Workers by Sector of Economic Activity – 1981, 1991, 2001

	1981		1991		2001	
Economic activity	Companies	Workers	Companies	Workers	Companies	Workers
Agriculture and fishing (a)	30.215	110.195	31.408	96.759	34.316	98.934
Extractive industry	4.477	56.791	3.617	46.360	3.837	36.164
Manufacturing industry	591.014	5.862.347	552.334	5.262.555	542.876	4.894.796
Energy, gas and water	1.398	42.878	1.273	172.339	1.983	128.287
Construction	290.105	1.193.356	332.995	1.337.725	515.777	1.529.146
Commerce and repair	1.282.844	3.053.706	1.280.044	3.250.564	1.230.731	3.147.776

	1981		1991		2001	
Economic activity	Companies	Workers	Companies	Workers	Companies	Workers
Hotel and civil services	212.858	644.223	217.628	725.481	244.540	850.674
Transport and communication	132.164	679.386	124.768	1.131.915	157.390	1.198.824
Credit and insurance	27.775	446.745	49.897	573.270	81.870	590.267
Other services	274.463	911.560	706.294	1.977.334	1.270.646	3.238.040
TOTAL	2.847.313	13.001.187	3.300.258	14.574.302	4.083.966	15.712.908

Although the number of workers as a whole grew over the twenty-year period 1981–2001, it is also evident that the number of employees in the manufacturing industry has remarkably decreased. The data are significant given that in the meantime the Italian population grew as a whole, albeit not at the pace of earlier decades. We can extrapolate backwards the area of investigation to uncover that until 1981 the number of people employed by the industry increased instead. A study by Margherita Russo and Elena Pirani (2006) that spans the half-century is useful. The tables, conveniently reconstructed and harmonized, show first the growth and then the fall of employment, both in absolute terms and in percentage.

Table 2 – Dynamics of Workers in Italy by Sector of Economic Activity, 1951–2001 (absolute values)

	1951	1961	1971	1981	1991	2001
Engineering	1.041.962	1.569.306	2.166.813	2.745.513	2.531.295	2.496.658
The rest of manufacturing	2.456.258	2.928.698	3.141.774	3.397.865	3.253.313	2.766.994
Services	100.802	110.194	170.550	702.928	1.147.988	2.208.853
Total economic activity	6.781.092	9.463.457	11.077.533	16.883.286	17.976.421	19.410.556
Total manufacturing	*3.498.220*	*4.498.004*	*5.308.587*	*6.143.378*	*5.784.608*	*5.263.652*

One could therefore think that – given the fall in the number of companies and workers during the twenty-year period from 1981 to 2001 – we have entered into a phase of deindustrialization. This is partly true (Gallino 2003), but the data relative to industrial output show that the fall in number of workers does not correspond to a fall in output. See on this matter the study by Menghini and Travaglia (2010) on the evolution of Italian industry, where the tables relative to the decade 1981–1991 (the eighties) and 1991–2001 (the nineties) show a noticeable increase in industrial output.

Intermediary surveys during the decade 2001–2011 are less 'linear' because of the two big epochal events that have characterized the 2000s: a) the terrorist attack on the USA and ensuing war in the Middle East; b) the major economic crisis that began in 2008 and is still with us. ISTAT data shows that in the 2008–2010 period, the slump in employment becomes much more pronounced, while industrial output also decreases. This happens in Italy as in other Western nations. However, the first data of the 9th census on industry and services (ISTAT 2011) confirm the trend of a decreasing number of industrial employees coupled with the growth of the total number of workers.

Table 3 – Comparison of the Dynamics of Workers in Manufacturing with Total Workers, 2001–2011

Type of data	Number of active enterprises		Number of workers	
Year	2001	2011	2001	2011
Total	4083966	4425950	15712908	16424086
Manufacturing industry	527155	422067	4810674	3891983
Other activities	3556811	4003883	10902234	12532103

In ten years, the number of blue-collar workers has decreased by approximately one million units. The following specific data are also quite significant:

Table 4 – Dynamics of Workers in Manufacturing and Repair of Computers and Machinery, 2001–2011

Type of data	Number of active enterprises		Number of workers	
Year	2001	2011	2001	2011
Manufacturing of computer and other electronic devices	5434	5693	139239	112055
Manufacturing of machinery and other equipment	21263	24584	451806	457956
Repair of computer and other domestic appliances	33659	26152	61512	46837

Here, we can clearly see that workers expelled from other manufacturing industries are not reabsorbed into the computing and machinery sectors. The number of enterprises active in the manufacturing of computers grew, while the number of workers in the same sector significantly shrank. We also notice a decrease of both enterprises and workers involved in computer repair. The only exception is the manufacturing of machinery, where we observe that the number of both enterprises and workers grow. But a growth of 6,150 workers compared with the loss of one million jobs can hardly be seen as evidence that workers made redundant by machineries are 'recycled' by the system as machinery constructors.

Summing up, notwithstanding turbulences connected to wars and financial crises, we can say that on the whole, during the last thirty years, a trend has emerged that is characterized by a fall in the number of industrial workers and an increase in industrial output. This should not astonish if one keeps in mind that productivity depends also on other factors. The other factor that grows noticeably during this same period is precisely automation, that is, the massive use of computers and robots in industrial manufacturing.

All this therefore leads one to think that there exists a relation between the fall of employment in industry and the growth of automation. This is the hypothesis we want to consider.

Allow us to open a bracket. It is known that data are not only read but also interpreted. A statistical correlation does not imply a causal dependence between the phenomena. Therefore the statistical data could just be a starting point, to which will later be added other elements, other considerations. But without statistics one goes nowhere. To those who say that statistics are unreliable and that one can therefore easily do without them, we reply with a well known popular diction: if money does not buy happiness, imagine misery. By analogy we say: if statistics do not give certainty, imagine mere impressions. Close brackets.

To start, we take into account the interpretation of Luciano Gallino, perhaps the greatest expert on the sociology of work and industry in Italy. We are trying to shed light first of all on the question of 'technological unemployment:'

> Technology is essentially a means to do two different things. On the one hand one may try to produce more, even much more, using the same amount of work. On the other hand one can try to use the potentiality of technology to reduce the workforce employed to produce a given volume of goods or services. And this leads to a very simple equation: as long as one manages to increase production, which means that as long as one manages to increase the markets, technology does not generate any unemployment because the work force remains constant and the only thing that grows are the markets. The markets however, different the one from the other, varied as they are, in general cannot expand forever. When the markets can no longer expand, technology is used mainly to reduce the workforce and then the spectre of technological unemployment begins to loom (Gallino 1999).

Economists tend to underestimate the problem of technological unemployment because one observes that, in percent, the unemployment due to the technological development of the last two hundred years has not been unsustainable. In general one avoids the issue saying that every new technology eliminates one job while creating another. Even if the computer takes an employee's job, there will be the need to build and maintain computers.

There is a grain of truth in this observation, but the issue is slightly more complex. This allegation always wafts the idea of the invisible hand, of the self-regulating market. In reality the system has so far had to thank the constant intervention of governments with policies of all kinds. The readjustment of the economic system, following the introduction of new and revolutionary technologies, does not happen in real time and without a price. If it is true that the worker or the employee that the machine has replaced can find another job, perhaps a new kind of job, it is also true that they might not have the skills required for the new job (for example: computer maintenance) and that, in order to acquire them they will need months and perhaps years – that is, if they are successful. Therefore the replacement work can arise one or two years after the work was lost. Humans are fragile machines – they do not survive more than a few days in the absence of a certain amount of calories, adequate clothing and a roof over their head. At the same time, these 'machines' tend to behave violently and disruptively if they come face to face with the prospect of their own destruction. Therefore even if it is the case that the market self-regulates, since it does not do so immediately, if one wants to avoid the instantaneous collateral effects of technological unemployment, one will have to play the public hand in addition to the invisible one.

This is what all governments have been doing, even those most liberal and capitalist. For over a century, governments have systematically obliged employers to reduce working hours, in order to compel them – against their interest – to maintain the number of employees.[15] They have instituted tools such as unemployment benefits or national insurance contributions, at times efficiently, at others creating pockets of parasitism. They have acquired the goods produced by the private sector via public contracts. They have funded retraining schemes and refresher courses for the chronically unemployed. And in the most tragic cases they have patched up the effects of economic crises by starting wars. On the one

---

15 Gallino makes the same observation: "In order to avoid reducing the working force and so to take too fast to the road of technological unemployment, one invented over a century ago the tools to reduce working hours. At one time, at the beginning of the [20th] century, one worked 3000 hours per year, in the middle of the century about 2500, and today most workers have a mean annual schedule of around 1600–1700 hours of work. This is one of the advantages of technology, that of being able to keep people employed while decreasing their performance."

hand conflicts reduce the population by sending entire generations to the front and on the other hand they allow the war industry to reabsorb the jobless. However cynical this may seem, it has happened and it continues to happen.

These tools, especially the systematic reduction of working hours and benefits to the temporarily unemployed, have so far worked rather well. Today the appearance of two new factors – globalization and artificial intelligence – has created a new situation with respect to the one generated by the first and the second industrial revolution. Globalization no longer allows one to operate on the basis of reduced working hours. To do so would be suicidal if not implemented at a global level and adopted by all nations. Although globalization has created *a single large market* it has not created a *single large society* led by a government that is its authentic expression. There probably exists a kind of 'shadow world government,' otherwise one cannot understand for whom or for what the national states are giving up their own sovereignty, but – if there is such a thing – it resembles a financial oligarchy that understandably defends its own interests, rather than an enlightened elite serving the interests of all. The idea that there could exist a market without a society has, as we can see, critical consequences.

In addition there is the matter of artificial intelligence. The idea that every job eliminated by a technology is sooner or later replaced by a job generated by that same technology is called into question by the nature of automation itself. Gallino (1999) writes: "This minor textbook equation prevails much less at the age of driven automation, the one that I call 'recursive automation.' The jobs that technology used to create soon after it had suppressed a certain number were partly recovered by the enlargement of the markets but partly also by producing technological means, that is, producing the same machines of goods and services that the markets absorbed up to a point. With automation applied to itself the machines produce other machines to automate, the process of automation attains very high levels and thus there is no longer any hope, or at least it is much reduced, to sooner or later find a new job in the sectors that produce the technology that eliminated the original job, the first job."

All empirical evidence shows that technological unemployment is more than a hypothesis. It is on the basis of data and graphs, and certainly not of moral principles, that sociologists criticize mainstream economics. Although it is dated, the book *Se tre millioni vi sembrano pochi*[16] is still instructive; in this analysis Gallino

---

16 This book has not been translated into English. The title means: *If three million seems not much. On the ways to fight unemployment.*

(1998) gives a central position to recursive automation, which until then had been given little heed. In the following review, sociologist Patrizio Di Nicola sums up the main ideas:

- To the myth that the upswing generates employment the author opposes Italian statistical evidence: in thirty years GNP has doubled, but the number of workers has only increased by 2,1%, that is of 400 000 units. But at the same time the number of resident citizens has increased by over 6 million;
- The idea that technology creates, long term, more jobs than it destroys was valid in the past, the author states, but is no longer so. The increase in productivity due to new machines can generate a positive occupational balance only if the markets absorb more merchandise. But in Italy companies function inside mature and partly static markets and export in these sectors is anything but easy;
- The advice to do 'like the Americans,' who seemed to have managed to create a phenomenal *job machine*, is founded on misleading presuppositions. In fact, on the one hand, the increase in the number of jobs is a direct consequence of the increase in the population (which between 1980 and 1995 went from 227,8 to 263,4 million units). On the other hand American job performance is aided by a somewhat relaxed statistical method. Count as employed the following: 6 million students aged 16–24, who nevertheless have been working for at least one hour in the week preceding the survey (maybe washed the neighbour's car or delivered the newspaper before going to college); 20 million *contingent workers*, people who work sporadically, when they can; 23 million part-time workers, who in reality correspond to 12 million fulltime jobs. And in the same way as they overestimate the number of employed – Gallino observes – the official statistics made in the USA underestimate the number of unemployed that, applying European criteria, ought to be over 12% instead of 5,3%. That is a little over the European mean.
- To the idea that what is most responsible for the poor levels of employment is the welfare state Gallino objects with some 'odd cases': Italy, with its 12,2% unemployed, spends 25,1% of GNP on welfare, while Holland, which has an unemployment rate of 6,5% spends the most: 29,8%. Denmark, the country where unemployment is the lowest in Europe, allocates 32,7% of GNP to social welfare. At the other extreme Spain, which invests less than we [Italy] in welfare, has a level of unemployment that is above 22%.

In brief, the growth of production and of productivity has not necessarily brought about the growth of employment. The relation between growth and employment is extremely weak in a country that does not produce technologies but imports them at best.[17] The American model is an illusion because the occupational data

---

17 "A country that mostly buys a technology researched and developed by other, increases its productivity, and therefore sees the number of jobs decrease but it does not see them recreated by this technology" (Gallino 2008, 17).

are 'inflated' (and, ten years after the publication of this book, the situation is even worse after the outbreak of the financial crisis). Welfare state – considering the graphs – rather than being an obstacle to growth appears to be a factor of production, but almost all Western countries tend to respond to the crisis by dismantling or reducing social benefits.

Not only that. One cannot even hope that someone who is expelled from industry will later necessarily be reabsorbed by the service sector (be it public or private), "because services are just as much susceptible to automation as the production of goods" (Bignami 2007). We will consider this aspect in greater detail.

2.3 Social stratification and new generation robots

Automation is already expanding beyond the manufacturing industry. The evolution of robots now has its effects also on the tertiary sector. In addition the presence of robots within the home is growing at the rate of 7–8% per year. According to the predictions of Bruno Siciliano, president of the International Society of Robotics and Automation, "out of the 66 billion dollars that will represent the cost of robotics in 2025, 35% will be that of personal and service robots" (Bignami 2007). This is why, if we have failed in the past to correctly formulate the social problem of robotization, it would be even more shortsighted not to formulate it now. "So from now on robots are everywhere. In our home, in the office, in our car. They take care of the elderly: South Korea has developed robots that control home electric appliances and remind the elder person when it is time to take his medicine. They serve as nurses to the sick (in the USA some prototypes are even taking their temperature) and they can also transform into tail-wagging puppies (the case of 'Aibo' among others); soon they will act as baby-sitters if it is true that some companies are researching how to 'teach' the automaton to rock a newborn."

Philosophers and scientists talk about all this, once in a while, in symposiums and conferences, but the question seems virtually absent from political agendas. The problem is underestimated for two main reasons: 1) the lobbying power of major industries, that only stand to benefit from robotization, and therefore feel no necessity to discuss the issue in wider terms, and 2) the widespread conviction that robots will never be able to imitate humans *all the way*. Yet, as Siciliano observes, today there are robots capable of doing the same work as a craftsman. "They are working in the zone between Vietri and Cava dei Tirreni where they are imitating the master potters." In practice the robot is not just able to imitate the assembly line production and surpass it in precision, but also that human imprecision of the craftsmen that is so characteristic of their product. An optical system records the craftsman's imprecise brush strokes, all different from one another. Using this information one writes a program which, when implemented in the robot, enables it to produce tiles that are all different from one another.

If we proceed in this way, robots could replace humans also for activities involving decision-making. Interviewed by Bignami (2007), Antonio Monopoli makes this forecast: "It is likely that with time one will produce robots with greater and greater ability to teach themselves. In fact we will have robots able to 'decide', a condition they share with humans." Once we have got that far, according to Bignami, "the expansion of robotics will also involve problems of ethics, and it is not fortuitous that one talks of 'Roboethics' at the ICRA conference. One problem that may arise is the possible inadequacy of the robot's response to events. In the case of injury, who would be responsible?" Monopoli replies that: "If the robot is regarded as a machine, the responsibility falls on the owner. But if the robot has a great capacity for self-learning and interaction with the external world, and the idea of robots working autonomously is socially accepted, one could not question the good intention of those who designed and commercialized the robot."

These problems are generally presented as having to do with roboethics, and therefore as *ethical* problems (that concern the whole of humanity and have to be solved with reference to universal principles) and not as chiefly *political* (that is, that concern the interests of a polis, a community, a faction, a social group). Now we should wish to stress that the problem – be it ethical or political – was born before, when the big industrial robots arrived to the factories. The robots' spread from the factories into homes and offices is if anything part of an *evolution* of the old problem that arose already with the industrial revolution. The ruling classes downgraded the problem of technological unemployment into a 'technical' one and certainly not 'ethical', as long as the victim of the process was the working class. It would be interesting if the same ruling classes were outraged should an anthropomorphic robot sit down at the desk of the CEO or an AI replace the manager in the control room of a multinational. If he were alive, Karl Marx would probably say that that the bourgeoisie wakes up to the ethical problem once the robot reveals itself able to replace also the manager, the artisan, the medical doctor, the teacher, when it acquires the ability to make decisions – and not only the proletarian at the assembly line. Again, the dominant group equates itself with *humanity* and turns its own political problem, its own class interests, into a universal ethical problem.

2.4 The need for a new socio-industrial policy

With the exception of radical ecologists and the supporters of degrowth, most people – regardless of which side of the political spectrum they are on – would claim that economic growth and a high rate of employment are good things. These are universally seen as goals that should be pursued. Let us therefore ask ourselves if the policies that the last several Italian governments have implemented are effectively rational – that is, do they allow these goals to be achieved. Current Italian political leadership seems to assume that growth and employment have no causal link with automation, given that this factor is repeatedly forgotten in analyses. Basically, politicians accept the thesis of the "Luddite fallacy" elaborated by economists.

Therefore, most policies have the aim of making the labour market more flexible or reducing the cost of labour. It is assumed that Italy would attract more investments, if it were easier for capitalists to fire workers and if workers were less 'choosy' when looking for a job. The policies based on these assumptions tend to create a favourable ground for brain drain and the immigration of unskilled workers. A mass of immigrants with reduced rights (given that they cannot vote and have temporary resident permits) is much more appealing to companies than skilled and demanding citizens.

Has this approach, adopted systematically since the early 1990s, produced positive results? Everything points to the contrary. As a matter of fact, the deregulation of the labour market and the gradual dismantling of the welfare state have not generated the expected results. Data from the World Economic Outlook Database of the International Monetary Fund (October 2012) show that, in the decade 2003–2013, Italy's growth has on the whole been −0.1%. This means that while the global economy keeps growing, Italy occupies one of the few positions with negative growth, together with Zimbabwe, San Marino, Greece, and Portugal (Aridas and Pasquali 2013).

These policies could be ineffective and, perhaps, even counterproductive exactly because the last generation of robots and computers have something to do with structural unemployment. On the one hand, highly automated industries, having not so many humans in the loop, are probably much more preoccupied about the cost of energy than the cost of labour. On the other hand, no matter how much the cost of labour and the rights of workers in a developed country can be reduced, low-tech industries will always find it more convenient to relocate to underdeveloped countries.

In other words, the very problem could be the postulate on which the system is built: its necessity – that is, the idea of the invariance of the mode of production. Hence, the solution to almost any contingent problem is primarily to patch it up (at low cost) to keep the system up and running for now – leaving the serious problems for future generations to sort out.

This is pretty obvious in the case of the politics of development and of social security policies. For decades Italian political leaders *spoken* of the necessity to stimulate scientific research, but talk remains always and only talk. In reality investment in research, both from the State and from private sources, is at its lowest.[18] Hence, it so happens that Italy – a nation belonging to the leading group of developed economies (of the G7 or G8) – has no manufacturers of computers or of mobile phones – to state two driving products of the new economical phase. The result is that technological development is certainly not slowing down, given that technology can also be imported. Rather the result is that one does not stimulate the sector that could reabsorb at least part of the technological unemployment.

As regards the politics of prevention, a now creaking system has been in place for a few decades, and it relies on two remedies: a massive immigration from the less developed countries and an increased age of retirement. The first remedy presupposes that there is an oversupply of jobs in Italy, while the second one shrinks the job market for the young – so this policy appears schizophrenic right from the start. Yet this policy is in fact the fruit of a plan that the Ministry of Work and Social Policies, under the leadership of Maurizio Sacconi, has put in black and white. If we read a document by the Directorate-General dated February 23rd 2011 with the title "Immigration for work in Italy" we discover that the Italian government feels the need to increase the number of immigrants: "In the period 2011–2015 the mean yearly requirement should lie around 100,000 while in the period 2016–2020 it should reach 260,000" (Polchi 2011). So in the next few years we will need to 'import' one million eight hundred thousand workers who would be added to the four million already residing in Italy (data from ISTAT)[19]. The conclusion that we will need six million

---

18 The Eurostat 2009 report on science, technology and innovation in Europe is unforgiving and positions Italy among the last. In 2007, the 27 member states invested in a total of less than 229 billion euros, or 1,85% of the European GNP. At the same time, the USA reached 2,67% of GNP, and Japan (in 2006) 3,40% of GNP. In Europe, only Sweden and Finland spent more than 3% (3,60% and 3,47% respectively), then there are 4 countries (Denmark, Germany, France and Austria) that spent over 2%. Italy invests little: 1,09% in 2001 and 1,13% in 2006. But it is data relative to employment that interests us most and these data are very discouraging. According to this report researchers in the EU represent 0,9% of employment, while in Italy they reach 0,6% (Eurostat 2009).
19 "Foreign residents in Italy on January 1st 2010 are 4.235.059, representing 7,0% of the total number of residents. On January 1st 2009 they represented 6,5%. During the year

immigrant workers in the next ten years derives from the following analysis: "The need for manpower is linked at once to job demand and job supply. On the side of supply one foresees that between 2010 and 2020 a decrease in the working population (employed plus unemployed) of 5,5% and 7,9%: from 24 million 970 thousand in 2010 it would fall to a value comprised between about 23 million 593 thousand and 23 million in 2020. On the side of demand the number employed would grow for a decade at a rate between 0,2% and 0.9%, reaching in 2020 23 million 257 thousand in the first case and 24 million 902 thousand in the second" (Polchi 2011).

Where is the error? For a start, one has not at all taken into account that we are not yet out of the crisis and that too many Italian companies, when they have not relocated, are closing down.[20] Among other things, now also 'historic' companies like FIAT threaten to relocate their production abroad. All this while brains are drained. And there is more. If what we have seen about automation is true, the calculation error is macroscopic. One cannot appraise employment on the basis of a presumed increase in production which, among other things, does not include a possible increase in productivity due to automation. Is it too much to ask of the Ministry of Work that they know what artificial intelligence is? If nurses and bricklayers will also be replaced by robots, what will then become of the six million immigrants that no one has really tried to integrate, but that have instead been regarded as stop-gaps to keep pension payments ticking over? What will six million people do – with different languages, religions and customs – when they have no home and no work, and, since they are not even citizens, will have no political rights and not be eligible for many kinds of social benefits? Has anyone ever asked if among these six million there is an even ratio of men and women (the required minimum to favour integration)? Has anyone ever asked what skills they have? If they can be given the jobs of the future? And how they feel about Italians? About Europeans?

Of course, one cannot blame only the centre-right government for this short-sighted policy, given that it is a bipartisan vision, where in fact some left-wingers would turn a blind eye to illegal immigration – and therefore not include them in the census nor ever attempt to plug the leak of the 'Italian system.' Even the

---

2009 the number of foreigners grew by 343,764 units (+8,8%), a very high increase, but lower to that of the two preceding years (494,000 in 2007 and 459,000 in 2008, +16,8% and +13,4% respectively), chiefly as an effect of fewer arrivals from Romania" (ISTAT 2010).

20 In 2010 in Italy there were over eleven thousand applications for bankruptcy – about thirty a day – representing an increase of 20% with respect to 2009 (Geroni 2011).

Catholics gloat at the government's document. Andrea Olivero, the national president of the ACLI, the Christian Association of Italian Workers, hurried to say that, "these data will expose the demagogy of those who go on about the threat of immigrants. Without them the nation would implode, and to welcome them civilly is not just a humanitarian act but also an intelligent strategy for the future (…). While the last few years have been dominated by an obtuse logic of containment that however has failed, we are happy that the Ministry of Work now looks realistically at the data because only then will it finally be possible to direct the government to the phenomenon of immigration that until now has been unsuccessful" (Polchi 2011).

A wise strategy for the future? Not at all, if the scenario analysis elaborated by futurist Hans Moravec in "The Age of Robots" (1993) is at least partly correct. According to him, in the first half of the 21st century "inexpensive but capable robots will displace human labor so broadly that the average workday would have to plummet to practically zero to keep everyone usefully employed." But since the possibility of reducing working hours is not even discussed, what we can expect is growing unemployment, or the growing precariousness of the labour market, or the creation of pointless jobs. Yet without going too far, it would be enough to familiarize oneself with Moore's Law, the rate of development of artificial intelligence, the prospect of robotics and nanotechnology, in order to understand that not many hands and perhaps even not many brains will be needed to maintain or augment the level of production.

The 'rough guess' planning by the Italian government leaves one therefore somewhat perplexed. If this is the vision of the future of the ruling class, then we should probably expect a gloomy scenario. The possible consequence of an underestimation of the new automation process could be "widespread immiseration, economic contraction and polarization between the wealthy, the shrinking working class and the structurally redundant" (Hughes 2004).

Actually something even worse may happen. It is unlikely that we will witness the peaceful extinction by starvation of humans replaced by AI in the production process. Before this takes place a revolt will break out. And perhaps this would even come as a surprise to some. Also Gallino (1999) states that we will have to expect social tensions. When asked if he foresees conflicts in the future he replies:

> Yes of course, even if these conflicts will be of various kinds. In the meantime the conflict we have now is due to growing inequalities. In all the industrial nations, including our own (and ours even to a lesser extent that the others) the technological development of the last 20 or 30 years has meant a high increase in inequality between the fifth that earns least and the fifth that earns most from their work. If you then consider the smallest percentages, the differences are even larger, above all in the United States, but also in nations like Great

Britain, France, our own, but even in China where inequalities have risen very much. This is a conflict that is as old as the world itself, but which nevertheless the technologies tend to accelerate and embitter. And then there are the conflicts that are, let us say, more intrinsically linked to the technologies. Many technologies meliorate life, allow one to work better, with less difficulty, many technologies entertain, they are intellectually stimulating, can serve as learning tools and so on. And then the difference that is introduced is that between those who can master these technologies, that give them a better life, and those who instead cannot make adequate use of them, either for economic reasons or for cultural reasons, perhaps also for political reasons. Let us not forget that in some states in the world the new technologies are subjected to censorship, limitations, police control and similar. Hence one of the major conflicts of the future will be between those who are full citizens, fully participating in the technological citadel, and those who instead have to camp outside its walls.

The conflict between the owner of the robots (the new means of production) and the unemployed who have been expelled from the processes of production (the new proletariat) is a looming menace on the horizon. Already a rate of unemployment of 10–12% creates social tensions and generates crime. Imagine what could happen if it reached a much higher rate. Obviously, it cannot be excluded that the present technological change is generating only temporary problems, like all the previous technological changes in the last two centuries. All our preoccupations could be dissolved by the birth of new jobs that we cannot even imagine. But we cannot also exclude the possibility that we might have to face a completely novel situation. The machines that will enter our society could be so intelligent that *almost all* human workers may soon become obsolete. We must also be prepared to face this scenario.

If this happens, if this is happening, the best solution is not banning AI, but rather implementing social policies that would permit us to have all the benefits of robotization and automation without the unwanted collateral effects of unemployment or increasing job precariousness. We must be ready to reactivate the policy of the gradual reduction of working hours and to introduce a citizen's income. We must be psychologically prepared to reverse the dominant economic paradigm. To revitalize the economy, we might not need people working harder. We might need people working less. "Working less means work for all" – as a notorious slogan states. We might need more holidays, more free time, more welfare state, more money to spend. These policies would certainly make human labour more expensive, but – contrary to what most economists think – this could be exactly what we need. The increase of the cost of labour makes "investments in automation increasingly attractive" (Hughes 2004), high-tech economies are more competitive than low-tech ones, more competitive economies can distribute better 'social dividends' to their citizens.

This is hard to see, if we divide the world into Luddites (those that want to ban the machines) and anti-Luddites (those that label a Luddite whoever dares to relate technology to unemployment), tertium non datur. A third way actually exists: one may want more robots, more computers, more intelligent machines, more technologies, *together* with a consistent change in the system apt to guarantee a rational and fair redistribution of wealth. As sociologist James Hughes (2004) put it: "It's time to make a choice: Luddism, barbarism or a universal basic income guarantee."

3. Citizens and Automata

3.1 Technology and unemployment

While discussing the role of slavery and the difference between instruments of production and instruments of action, Aristotle (1995, 5) states that "a slave is a sort of living possession, and every assistant is like a superior tool among tools. (For if each tool could perform its own task either at our bidding or anticipating it, and if – as they say of the statues made by Daedalus or the tripods of Hephaestus, of which the poet says, 'self-moved they enter the assembly of the gods' – shuttles shuttled to and fro of their own accord, and pluckers played lyres, then master-craftsmen would have no need of assistants nor masters any need of slaves)." In other words, if automata were sophisticated enough to replace humans in every activity, slavery and work would be unnecessary.

Analysing the motives behind the revolt of the Luddites, in 18th and 19th century Europe, Karl Marx (1976, 532–3) made a sarcastic comment on Aristotle's philosophy of technology: "Oh! those heathens! They understood nothing of Political Economy and Christianity (…). They did not, for example, comprehend that machinery is the surest means of lengthening the working-day. They may perhaps have excused the slavery of one person as a means to the full human development of another. But they lacked the specifically Christian qualities which would have enabled them to preach the slavery of the masses in order that a few crude and half-educated parvenus might become 'eminent spinners,' 'extensive sausage-makers' and 'influential shoe-black dealers.'"

Both the Luddites and Marx have noticed that machinery did not free humans from labour, but rather caused unemployment and the inhumane exploitation of those still employed. However, they proposed different remedies. As is well known, the Luddites saw the solution in the destruction of the machines,[21] while Marx and the socialists preached that the proletarians would benefit more from a revolution aimed at taking full possession of the machines (the means of production). It is worth noting that not only the anti-capitalist front, but also a supporter

---

21 According to David F. Noble (1995, 3–23) the Luddites did not destroy machines because of technophobia, but because of necessity. They had to choose between starvation, violence against the capitalists, or property destruction. The last choice was the most moderate way to protest against unemployment and the lack of compassion of the factory owners.

of free market economy like John Stuart Mill (1848) was honest enough to admit that "it is questionable if all the mechanical inventions yet made have lightened the day's toil of any human being."

Since then, it has been restlessly debated as to whether technological development really frees humans from work, or on the contrary produces more exploitation and unemployment. There is a copious literature supporting the first or the second thesis, spanning over the last two centuries. And the debate is still going on.

The theory that technological change may produce structural unemployment has been repeatedly rejected by neoclassical economists as nonsense and labelled 'the Luddite fallacy.' These scholars contend that workers may be expelled from a company or a sector, but sooner or later they will be hired by other companies or reabsorbed by a different economic sector.

It is however well known that economics is a multi-paradigmatic discipline. Therefore, supporters of the idea of technological unemployment keep appearing on the stage. In the 1990s, right after the beginning of the Internet era, a few influential books focusing on the problems of automation and artificial intelligence appeared. Among these, we may cite books like *Progress without People* by David F. Noble (1995), *The End of Work* by Jeremy Rifkin (1995), or *Turning Point* by Robert U. Ayres (1998).

Noble stands in "defence of Luddism" and moves accusations of irrationalism to "the religion of technology" on which modern society is supposedly based. According to him, "in the wake of five decades of information revolution, people are now working longer hours, under worsening conditions, with greater anxiety and stress, less skills, less security, less power, less benefits, and less pay. Information technology has clearly been developed and used during these years to deskill, discipline, and displace human labour in a global speed-up of unprecedented proportions" (Noble 1995: XI).

Rifkin points out that people that lose a low-skilled job often lose the only job they are able to do. Many of people involved for instance in assembly or packaging can barely read and write. They are on the lowest rung of ability and learning. However, the new job that arises from the machine that 'steals' their job is one involving taking care of that machine, which often requires high school computer programming, if not a college degree in computer science. These are in turn qualifications requiring abilities at the higher end of the ladder. In brief, "it is naive to believe that large numbers of unskilled and skilled blue and white collar workers will be retrained to be physicists, computer scientists, high-level technicians, molecular biologists, business consultants, lawyers, accountants, and the like" (Rifkin 1995, 36).

Finally, Ayres emphasizes the fact that, even if we admit that workers can be relocated, new jobs may be less satisfactory than old jobs, in terms of wages, fulfilment, and security. And this is not an irrelevant aspect. This is evidence that globalization and automation are good for some social classes and bad for others. Indeed, "many mainstream economists believe that in a competitive free market equilibrium there would be no unemployment, since labor markets – like other markets – would automatically clear. This means that everyone wanting a job would find one – at some wage." The problem is that "there is nothing in the theory to guarantee that the market-clearing wage is one that would support a family, or even an individual, above the poverty level" (Ayres 1998, 96).

The reaction to these works is in the same vein as criticism of previous Luddite predictions. The main argument against the thesis that automation produces structural unemployment is that the many times predicted catastrophes never happened. The rate of unemployment may go up and down, but it never happened that technological change has produced an irreversible crisis. Ten years ago, Alex Tabarrok (2003) confessed to be "increasingly annoyed with people who argue that the dark side of productivity growth is unemployment." He added that "the 'dark side' of productivity is merely another form of the Luddite fallacy – the idea that new technology destroys jobs. If the Luddite fallacy were true we would all be out of work because productivity has been increasing for two centuries."

Apparently this is an invincible argument, but it did not stop predictions of technological unemployment. The reason is simple: Tabarrok reaches his conclusion by means of inductive reasoning. The premises of an inductive logical argument provide some degree of support for the conclusion, but do not entail it. That is, the fact that the catastrophe did not happen until now does not imply that it cannot happen today or tomorrow. After all, every technological change is qualitatively different from the previous ones. In particular, the novelty of the present situation is that artificial intelligence and its products (computers, robots, industry automation, Internet, etc.) intertwine with globalization – AI unfolds in a situation in which nation-states have a limited possibility to implement corrective policies. There is also a suspicion that not only the speculations of the bankers but also accelerating computer technology have contributed to the genesis of the financial crisis which exploded in September 2008 with the bankruptcy of Lehman Brothers. This is, for instance, the point made by Martin Ford in his *The Lights in the Tunnel* (2009).

On 13th June 2013, Nobel Prize winner Paul Krugman added his voice to this debate with an article significantly entitled "Sympathy for the Luddites." This economist recognizes that, in the past, the painful problems generated by mechanization were solved thanks to the more intensive education. However, the problems generated by artificial intelligence are not solvable the same way, because they effect educated workers as well. Thus, today, "a much darker picture of the effects of technology on labor is emerging." Krugman reminds us that

> The McKinsey Global Institute recently released a report on a dozen major new technologies that it considers likely to be 'disruptive,' upsetting existing market and social arrangements. Even a quick scan of the report's list suggests that some of the victims of disruption will be workers who are currently considered highly skilled, and who invested a lot of time and money in acquiring those skills. For example, the report suggests that we're going to be seeing a lot of 'automation of knowledge work,' with software doing things that used to require college graduates. Advanced robotics could further diminish employment in manufacturing, but it could also replace some medical professionals.

In the present investigation, we will tentatively assume that the picture drawn by Krugman and others is correct, and we will try to extrapolate possible futures from it. The debate seems to be mainly crystallized on the dichotomy "technology is bad" (Luddites, technophobes) versus "technology is good" (anti-Luddites, technophiles), but it is worth noting that there are many more armies on the battlefield. As we have seen above, Marx built his own value judgment by taking into account one more variable: the system. In short, his position was "technology is good, the system is bad." This third position disappeared somehow in the shadow in the second half of the 20th century, for many reasons that I cannot discuss here, but it seems to be an indispensable one. One does not need to be a revolutionary socialist in order to ask for a more complex analytical model. Krugman (2013) also points his finger at the degeneration of the system, more than at technology itself. The Nobel Prize winner stresses that

> the nature of rising inequality in America changed around 2000. Until then, it was all about worker versus worker; the distribution of income between labor and capital – between wages and profits, if you like – had been stable for decades. Since then, however, labor's share of the pie has fallen sharply. As it turns out, this is not a uniquely American phenomenon. A new report from the International Labor Organization points out that the same thing has been happening in many other countries, which is what you'd expect to see if global technological trends were turning against workers.

As a response, he does not propose to get rid of the machines, but to activate a policy of redistribution of wealth, "one that guarantees not just health care but a minimum income, too." Note that he is not asking for a radical change of the system, as Marx does, but just to fix it. Therefore, it is important to elaborate a model capable of taking into account positions with a focus on the system, and of different types, like those of Krugman or Marx.

3.2 Some methodological tools for scenario analysis

Many futurological speculations follow a simple pattern: they always and invariably see technology as a cause and social structure as a consequence, never the other way round. Therefore, the attitude toward technology becomes the one that really matters.

In other words, these theories do not give much weight to the role that social and industrial policies can play in shaping the future. This typically happens when futurologists are also engineers. They know better than anybody else how technologies are produced and work, but they also tend to underestimate the complexity of the social, political, and economic world. On the contrary, social scientists teach us to view social problems in a more complex way, to be aware that it is often hard to distinguish cause and effect, and that the forecasts themselves sometimes bring about the very process being predicted – the so-called 'self-fulfilling prophecy' (Merton 1968, 477). In the social reality, one more often observes a chaotic interaction between different variables, rather than a simple cause and effect chain.

The society of the future will partly depend on structures inherited from the past that cannot be easily changed. Some of our behaviors depend on what sociologists call 'social constraint,' on what philosophers call 'human condition,' and on what biologists call 'human bio-physical constitution' – all variables that change very slowly.

However, the future will be partly shaped also by crucial decisions of powerful people and by the predictions of influential futurologists. Even if the different attitudes and beliefs of individuals (Terence McKenna would call it 'the cultural operating system') can be rather stable and randomly distributed in society, the equilibrium of power may change in a sudden and unpredictable way. The rulers may become the ruled. Marginal worldviews may become mainstream ideas. So, what really matters is the attitudes and beliefs of the ruling class (politicians, bankers, entrepreneurs, top managers, scientists, opinion leaders, etc.) *in the moment* in which crucial decisions have to be taken. That is why, to draw pictures of possible futures we need models (attitudinal typologies) a little more complex than a simple dichotomy 'technophobes vs. technophiles.' To start we propose an attitudinal typology that combines 'technological growth'[22] and 'the system.'

---

22 See the definition of 'technological growth' by EconModel (www.econmodel.com): "Economic growth models (the Solow growth model, for example) often incorporate effects of technological progress on the production function. In the context of the Cobb-Douglas production function $Y = K^a(L)^{1-a}$, we can identify three basic cases:

We will call 'growthism' a positive attitude to technological growth and 'degrowthism' its antonym. We will call 'conservatives' those that support the invariance of the system and 'revolutionaries' those that want to change it.

1. *Attitudinal Typology toward 'Technological Growth' and 'The system'*

		The system	
		Accept	*Reject*
Technological growth	*Accept*	Conservative Growthism	Revolutionary Growthism
	Reject	Conservative Degrowthism	Revolutionary Degrowthism

In order to talk about a truly revolutionary 'system change' we will stipulate here that at least points 5 and 7 of Marx and Engels's *Manifesto* (1948) must be fulfilled, namely: a) "Centralization of credit in the hands of the state, by means of a national bank with State capital and an exclusive monopoly"; b) "Extension of factories and instruments of production owned by the State." As a consequence, we would say that a system change (or revolution) had occurred in EU and USA, if and only if the two central banks (BCE and FED) were nationalized and the robotized industries of the two countries were owned by all citizens.

When it comes to promoting or opposing technological growth, we may find different perspectives: some believe that technology develops in a spontaneous way, while others believe that governments (even in capitalist countries[23]) play a crucial role in shaping science and technology by mean of industrial policies. Obviously, as often happens, the truth is somewhere in the middle. We may find historical examples supporting the first or the second idea. It is however important what the ruling class believes, and here again we may have various combinations between attitudes.

---

labor-augmenting (or Harrod-neutral) technological change $Y = K^a(AL)^{1-a}$, capital-augmenting technological change $Y = (AK)^a L^{1-a}$, and Hicks-neutral technological change $Y = AK^a L^{1-a}$."

23 US science and technology would probably be much different if the government did not have so many contracts with the military-industrial complex.

2. *Attitudinal Typology toward 'Technological growth' and 'Industrial policies'*

		Industrial policies	
		Accept	*Reject*
Technological Growth	*Accept*	Programmed Growthism	Spontaneous Growthism
	Reject	Programmed Degrowthism	Spontaneous Degrowthism

Finally, when we shift attention from technology to people, we may find other combinations of attitudes. Among growthers, some are happy with the market distribution of wealth, while others call for the social redistribution of wealth. The same divide may appear also among degrowthers. It is important to emphasize, once again, that social policies do not necessarily imply a system change.

3. *Attitudinal Typology toward 'Technological growth' and 'Social policies'*

		Social policies	
		Accept	*Reject*
Technological Growth	*Accept*	Redistributive Growthism	Distributive Growthism
	Reject	Redistributive Degrowthism	Distributive Degrowthism

By using these models, we will try to speculate what could happen in the near future to an industrialized nation-state. We will not explore all the possible combinations of the above presented attitudes, but only four possible scenarios. We will examine the possibilities of a *future* to some extent shaped by the desiderata of the ruling class (planned), although with different orientations, involving the *extinction* either of workers or robots. Then we will consider the same two outcomes, but as unwanted (unplanned) consequences of other attitudes and policies.

4. *Typology of possible future scenarios*

		Future	
		Unplanned	*Planned*
Extinction	*Workers*	Unplanned end of work scenario	Planned end of work scenario
	Robots	Unplanned end of robots scenario	Planned end of robots scenario

3.3 The unplanned end of work scenario

The unplanned end of work scenario is generated by 1) technological growth; as an outcome of 2) the political and economic system not changing; 3) spontaneous growth; and 4) market distribution of wealth. Let us see how.

One author who has tried to foresee the possible developments of automation society on the basis of these premises is Hans Moravec. As a robotic engineer he has a solid grounding in technology, from which he extrapolates present data and projects them into the future. Moravec offers a very interesting point of view that it is worth thinking carefully about. He shows us what might happen in the case of *laissez-faire*, that is, if governments do not try to guide the course of future history.

In part one of the essay "The Age of Robots," Moravec (1993) describes four generations of universal robots which happen to coincide with the first four decades of the 21st century. We will not enter into the technical details, but limit ourselves to observe that the first generation of robots is that of robots that we sometimes see on television or in exhibitions, while the second generation is already able to replace humans in a variety of tasks outside manufacturing; the third generation displays even more 'human' traits and therefore competes with human labor in all sectors, while the fourth displays traits that are downright 'superhuman.'[24]

In the second part of the article, Moravec dwells on the social consequences of the emergence of universal robots, distinguishing short, mean and long term consequences. In the long term, according to Moravec, the superhuman robot will gradually be able to design even more potent and intelligent 'children,' and thus robots will acquire traits that are 'semi-divine.' Machines will merge with those humans who stay on – via the technology of mind-uploading – and will colonize space, converting other inorganic matter into thinking matter. These are bold speculations, but not at all impossible. We leave them however to the reader's curiosity.

Here it is enough to analyse the short term, which coincides with the first half of the 21st century. Moravec – who is anything but a Luddite or a left-wing

---

24 Moravec writes: "In the decades while the 'bottom-up' evolution of robots is slowly transferring the perceptual and motor faculties of human beings into machinery, the conventional Artificial Intelligence industry will be perfecting the mechanization of reasoning. Since today's programs already match human beings in some areas, those of 40 years from now, running on computers a million times faster as today's, should be quite superhuman."

extremist[25] – recalls first of all the painful transition from the agricultural society to the industrial society. The human cost of millions of workers forced to cram in the suburban areas of industrial districts and to compete for badly paid jobs that were never enough to satisfy demand. This period brought child labor, precarious employment, and inhuman working hours without any social security, health care, trade unions, or pension schemes. We exited this 19th century 'savage capitalism' because of trade unionization, revolutions and reforms, to finally arrive at the welfare state. In particular, the system has been saved thanks to the recurring reduction of working hours, aimed at counteracting technological unemployment and reducing exploitation. But in the era of robots will it be possible to continue with these reforms?

Not according to Moravec because even if working hours continue to fall (which it should be said is no longer even the case), their decline "cannot be the final answer to rising productivity. In the next century inexpensive but capable robots will displace human labour so broadly that the average workday would have to plummet to practically zero to keep everyone usefully employed." While governments may oblige private companies to reduce the working time of employees, they certainly can't oblige companies to hire and pay people to do nothing. But this is not the only problem. Even today, many workers are re-employed to perform 'frivolous' services and this will be even truer in the future, because as we have seen, services requiring efficiency rather than creativity will also be performed by robots. In practice, the function of humans is and will increasingly be to 'entertain' other humans with games, sports, artistic works or speculative writings (like this one). Some people are even paid to do trivial and utterly uninteresting jobs: think of some state-employed bureaucrats, often hired for no other purpose than an attempt to reduce unemployment, assigned to useless if not downright harmful tasks of control and regulation, who therefore end up as a burden to other citizens.

> Will we all be assigned to frivolous or useless services? It could be one solution, but it would seem that even this road is blocked. The 'service economy' functions today because many humans willing to buy services work in the primary industries, and so return money to the service providers, who in turn use it to buy life's essentials. As the pool of humans in the primary industries evaporates, the return channel chokes off; efficient, no-nonsense robots will not engage in frivolous consumption. Money will accumulate in the industries, enriching any people still remaining there, and become scarce among the service providers. Prices for primary products will plummet, reflecting both the reduced

---

25 John Horgan describes Hans Moravec as a Republican 'at heart,' a social Darwinist and a defender of capitalism, in *The End of Science* (Horgan 1997, 255).

costs of production, and the reduced means of the consumers. In the ridiculous extreme, no money would flow back, and the robots would fill warehouses with essential goods which the human consumers could not buy.

If we do not reach this extreme, there will in any case be a minority of capitalists (the stockholders) continuing to make profits thanks to a legion of efficient workers who do not go on strike, do not fall ill, work twenty-four seven, demand a 'salary' equal to the cost of energy and, to cap it all, need no pension because they will retire to a landfill. While for the mass of workers employed at frivolous services or at transmitting information (the so called knowledge industry) and for the chronically unemployed (the proletariat) the prospect is a return to the Middle Ages. Moravec effectively reminds us that "an analogous situation existed in classical and feudal times, where an impoverished, overworked majority of slaves or serfs played the role of robots, and land ownership played the role of capital. In between the serfs and the lords, a working population struggled to make a living from secondary sources, often by performing services for the privileged."

A rather discouraging scenario. And very disturbing, if one keeps in mind that it is an enthusiastic robotic engineer and supporter of capitalism who predicted it. In reality Moravec – perhaps worried by the apocalyptic scenario just outlined – hastily adds that things may not go this way. That is, he envisages an alternative scenario, the possibility of a different future that nevertheless implies a new awareness and an attempt to have history take another path.

We will not necessarily venture back to the Middle Ages because today's workers have reached such a level of political awareness and education that they would hardly allow a minority of capitalists to reduce them to slavery. Were we to arrive at such a level of degradation the people would "vote to change the system." But this choice implies a different scenario, a planned one.

3.4 The planned end of robots scenario

The planned end of robots scenario is generated by 1) technological degrowth; as an outcome of 2) a radical change in the political and economic system; 3) programmed degrowth; and 4) social redistribution of wealth. Let us see how.

To remedy the evaporation of humans from the working environment, various types of solutions have been proposed. Faced with a 'technological apocalypse' many are tempted by the idea of a return to the past. More and more citizens appear fascinated by the perspective of a degrowth in technology and industry – and not only visceral technophobes like the 'Unabomber' Theodor Kaczynski. Therefore it would seem that we must include also this idea in our discussion, although no political agenda is currently contemplating a ban on artificial intelligence. Supporters of this position have been given various labels: Luddites, primitivists, passéists, retrograders, reactionaries, bioconservatives, radical ecologists, etc. Since the idea finds consensus both on the left and right, even though its most radical version has as yet no representatives in Parliament, we have decided to call its supporters 'degrowthers' – a term that does not yet have strong political connotations and that therefore lends itself to this technical use. By symmetry we call 'growthers' the supporters of limitless growth (scientific, technological, industrial, economical).

First of all it must be stressed that the degrowthist idea is rather simple and forthright. Its simplest formulation does not demand any particular intellectual effort, any particular competence, but rather a gut reaction: "If technology is bad, ban it!" The message is simple, clear and limpid. For this reason, it has had some success in the media.

A slightly more careful analysis shows however that giving up technologies based on artificial intelligence carries no fewer risks than does their diffusion inside a framework of *laissez faire*. Indeed a policy of degrowth, that is, one geared for maintaining or restoring obsolete systems of production, would not allow the country that adopted it to stand up to the competition of other nations in a global economy. At the level of quality and prices, goods produced by artisanship would not withstand the competition of those produced by a mixed human-robotic system or even an altogether robotic one. Therefore, were one to ban AI, unemployment would not even be re-absorbed in the short term. Not only would unemployment not disappear, but the worsening of other economic parameters and the collapse of many companies would likely increase it.

Obviously, all degrowthists are not naïve and therefore we should expect a second policy to take place at the same time as a ban on AI: economic autarchy. It is not by chance that degrowthists are generally also anti-globalization. Leaving the global market would end the competition between national and foreign goods and services, and employment could thus be rescued.

This argument may seem sensible when formulated this way, but it too would carry a hefty bill. Exiting the global economy, closing the borders, and imposing a duty on imports, would rescue the situation in the short run by creating a kind of poor but self-sufficient economic enclave. In the long-term however this economy would be under the constant threat of a black market of technologically advanced products from abroad. Repression by the police or the military would be necessary to counteract internal mafias that, via smuggling, would look after their own interests and those of foreign companies. The repression could however convince the same mafias, or foreign governments serving large corporations, to stir up rebellions inside the autarchic system. In other words, a system at once autarchic and degrowthist – given its technological weakness – would make itself vulnerable to being swept away at any time by systems that are technologically more advanced, via conventional and non-conventional wars. This scenario should be kept in mind, unless one has unconditional faith in human beings and thinks of them as capable only of intentions that are benevolent.

The third move a degrowthist party could make in order to avoid having this sword of Damocles above its head is that of as attempting to impose a global ban on artificial intelligence. This is a clearly utopian vision, because an agreement between most sovereign states would not be enough. Just a few dissident growth-oriented nations would be suffice to nullify the contract. The realization of this utopia would require a global degrowthist empire which is a practical impossibility. Such a project can only be imposed by a global hegemon, and any country that rejects on principle the most revolutionary and powerful technologies cannot remain a global hegemon. It is often said that science-fiction ideas are the prerogative of technophilic futurologists, but in reality the idea of a global ban on advanced technologies is the most 'science-fictive' idea of all. We do however want to continue to examine this hypothesis for the sake of discussion.

Let us suppose then that, by some sort of miracle, something like a degrowthist empire came to be (maybe as a result of the global hegemony of a degrowthist religion). So now the question is: how long can it last? This global political regime must not just do away with computers and robots but also with the whole of science that allows the realization of these machines, that is, with the *know-how*. The degrowthists must destroy universities and libraries, burn books and journals, destroy data banks, arrest or physically eliminate millions of scientists and engineers who might revitalize AI, as well as all growthist citizens who might side along them. Should anything escape the degrowthist thought police, or once the 'purification' terminated, should bright and curious children be born who were able to revitalize science, this would be a U turn. A clandestine growthist movement and a black market would be born. The police state would find itself having to fight with obsolete means hypertechnological dissident guerrilla groups. It is hard to imagine that the new system would not sooner or later be defeated by these groups.

It is more than rhetoric when we say that "the world *must* move forward" or "you cannot stop the clock." It is also an acknowledgement of two social dynamics that do not allow growth and progress to stop for good. These two elements are *the will to power* – a force that moves human history or, in the sense asserted by Friedrich Nietzsche, the life of the universe itself – together with Francis Bacon's simple observation that *technology is power* (*scientia potentia est*). In other words, the drive for technological mastery cannot be suppressed, and degrowthist victories are always temporary. This happened for example when Judeo-Christianity overcame – with the complicity of other catastrophic events like Barbarian invasions, natural catastrophes and epidemics – the thousand-year old Graeco-Roman civilization. All that was needed was to leave lying around bits and pieces of that great philosophical, scientific, artistic, technological, commercial and military civilization for its spores to wake up and regenerate under other forms, despite the severity and conscientiousness of the Inquisition (Russo 2004, Pellicani 2007, Campa 2010a).

Therefore, it does seem that the degrowthist solution, in addition to being inefficient and risky, is above all impracticable in its extreme forms. It is not by chance that the governments of the developed world have until now tried to remedy the problem of technological unemployment with all the means save one: banning the new technologies. This however does not mean that technological degrowth is impossible. Actually, industries and technologies can disappear in some regions of the world even if they are welcome. This is another scenario that deserves to be explored.

3.5 The unplanned end of robots scenario

The unplanned end of robots is generated by 1) technological degrowth; as an outcome of 2) the political and economic system not changing; 3) spontaneous degrowth; and 4) either the market distribution *or* the social redistribution of wealth. Let us see how.

If we look at the programs by parties represented in most Western Parliaments, be it the governing coalition or the opposition, we discover that they are all more or less favourable to growth. It is rare to find a member of Parliament who waves the flag of technological or economic degrowth. At most we find politicians who, courting degrowthist votes, speak of 'sustainable growth.' Similarly, we do not find anyone who welcomes upheavals, social conflicts, high unemployment rates and widespread crime. The ideal societies of the various political forces differ in some essential aspects (some dream of a Christian society, others of a secular one, some want it to be egalitarian, others meritocratic, and so on). But as far as growth and employment are concerned, they all agree – at least in principle – that these are positive. Even those who want to abolish the democratic-capitalist system (the political forces at the two extremes: fascists and communists) and who therefore do not rule out a phase of social conflict, do not dream of a permanent chaos, a society of temps, jobless, sick, poor and criminals. They view their ideal society as one fulfilling material needs, offering spiritual harmony and possibly without crime. They want to go beyond capitalism precisely because, in their opinion, it fails to guarantee all this.

However, we can still find political forces that cause degrowth or social pathologies, out of incapacity, corruption or short sightedness. The worry that traditional political parties lack a vision of the future and that this may generate social instability has been expressed by several social scientists.

For instance, now, Europe is facing a very delicate political and economic crisis, characterized by economic depression and a high rate of unemployment. As a response, the EU has been imposing an austerity policy on member states to reduce budget deficits. The European ruling class seems to be firmly convinced that the best recipe for stimulating economic growth is the deregulation of the labour market and the reduction of government spending. This perspective has been criticized by many economists. For instance, Boyer (2012) predicts that this economic policy will fail, because it is founded on four fallacies. First, the diagnosis is false: the present crisis is not the outcome of lax public spending policy, but "it is actually the outcome of a private credit-led speculative boom." Second, it is fallacious to assume "the possibility or even the generality of the so-called 'expansionary fiscal contractions.'"[26] Third, it is wrong to assume that a single policy may work for all states: "Greece and Portugal cannot replicate the hard-won German success. Their productive, institutional and political configurations differ drastically and, thus, they require different policies." Fourth, "the spill-over from one country to another may resuscitate the inefficient and politically risky 'beggar my neighbour' policies from the interwar period."

The analysis produced by Boyer is quite convincing. However, if it is true that unemployment is partly due to growing and evolving automation, the dichotomy austerity versus government spending, or neoclassical economics versus Keynesian theory, is simply inadequate to draw a complete picture of the situation. It is missing one of the main points.

A similar problem can be observed also in the USA. Sociologist James Hughes (2004) noticed as long as ten years ago that in the USA "newspapers are full of

---

26 The Expansionary Fiscal Contraction (EFC) hypothesis was introduced by Francesco Giavazzi and Marco Pagano (1990), by using the fiscal restructurings of Denmark and Ireland as examples. They summed up their argument as follows: "According to conventional wisdom, a fiscal consolidation is likely to contract real aggregate demand. It has often been argued, however, that this conclusion is misleading as it neglects the role of expectations of future policy: if the fiscal consolidation is read by the private sector as a signal that the share of government spending in GDP is being permanently reduced, households will revise upwards their estimate of their permanent income, and will raise current and planned consumption."

bewildered economists scratching their head at the emerging jobless recovery. The right reassures us that job growth is right around the corner, although it wouldn't hurt to have more tax cuts, deregulation, freer trade and lower minimum wages. Liberals counter that we can cut unemployment with more job retraining, free higher education, more protectionism, more demand-side tax stimulus and non-military public sector investments." According to this sociologist, "the problem is that none of these policies can reverse the emerging structural unemployment resulting from automation and globalization."

We have seen that, in certain conditions, the neoclassical politico-economic approach may lead to the unplanned end of work scenario imagined by Moravec. However, the same approach, in the presence of a new great depression (an hypothesis that Moravec did not consider in 1993), may lead to an end of robots scenario. Austerity policies and a bad application of neoclassical principles are indeed already producing the deindustrialization of some countries (e. g. Italy).

However, the Keynesian approach *per se* may also lead to the unplanned end of robots scenario. To explore this possibility, all we have to do is to analyse the first path indicated by Moravec to escape the gloomy perspective of an unplanned end of work. The obligatory path is to push through with the politics of the gradual reduction of working hours while preserving people's purchasing power. This could take place in two different ways: 1) by redistributing income via taxation; or 2) by re-distributing shares in corporations and thereby also their profits. In each case people are excluded almost entirely from the production loop, but the first path may have a collateral unwanted effect.

Through the redistribution of income via taxation, the circulation of money can be reactivated by governments as soon as it slows. In this case citizens' incomes would be equal or similar, but in any case sufficient to keep production going via consumption. However, since the level of taxation would be decided by the people, the system could collapse should this level become unsustainable in a still competitive system. In other words, too heavily taxed robotic industries would fail, leaving the whole population without an income. Without support for an expansion of public ownership governments may be not able to help or buy failing industries. If central banks remain private, governments do not control the money supply and will have to finance themselves on secondary markets as if they were private companies. This situation could render it impossible to implement an effective industrial policy and would precipitate the nation-state in a vicious circle leading to deindustrialization and, therefore, unwanted derobotization.

3.6 The planned end of work scenario

The planned end of work scenario is generated by: 1) technological growth; as an outcome of 2) a radical change in the political and economic system; 3) programmed growth; and 4) the social redistribution of wealth. Let us see how.

To explore this scenario, we have to follow the second path suggested by Moravec. This path is a kind of socialist-capitalist hybrid founded on allowing the population to own the robotic industries by giving part of the shares to each citizen at birth. In this case, incomes would vary with the performance of the companies. Therefore, for people, it would become more important to elect the best top managers for their factories, instead of the best members of Parliament. Everybody would have enough to live off, but salaries could no longer be decided by political votes. Even if this solution preserved a few features of capitalism (competition and a market economy), the change it entails would be more systemic than what it appears *prima facie*. True, Moravec does not discuss at all the problem of the banking system and the control of money, but to assign at least the property of the productive system directly to the citizens is more 'socialistic' than taxing the rich to give some charity to the poor. We quote Moravec's passage in its entirety:

> The trend in the social democracies has been to equalize income by raising the standards of the poorest as high as the economy can bear–in the age of robots, that minimum will be very high. In the early 1980s James Albus, head of the automation division of the then National Bureau of Standards, suggested that the negative effects of total automation could be avoided by giving all citizens stock in trusts that owned automated industries, making everyone a capitalist. Those who chose to squander their birthright could work for others, but most would simply live off their stock income. Even today, the public indirectly owns a majority of the capital in the country, through compounding private pension funds. In the United States, universal coverage could be achieved through the social security system. Social security was originally presented as a pension fund that accumulated wages for retirement, but in practice it transfers income from workers to retirees. The system will probably be subsidized from general taxes in coming decades, when too few workers are available to support the post World War II 'baby boom.' Incremental expansion of such a subsidy would let money from robot industries, collected as corporate taxes, be returned to the general population as pension payments. By gradually lowering the retirement age towards birth, most of the population would eventually be supported. The money could be distributed under other names, but calling it a pension is meaningful symbolism: we are describing the long, comfortable retirement of the entire original-model human race.

Moravec is crediting the idea of the end of work to engineer James Albus, but it is rather a multiple discovery. For instance, James Hughes (2004) also comes to a similar conclusion, even if he would probably implement the distribution of wealth in a different way than Albus. In any case, he warns that "without a clear strategic goal of a humanity freed from work through the gradual expansion of automation and the social wage, all policies short of Luddite bans on new technology will have disappointing and perverse effects. If liberals and the left do not re-embrace the end of work and the need to give everyone income as a right of citizenship, unconnected to employment, they will help usher in a much bleaker future of growing class polarization and widespread immiseration. If libertarians and the right do not adapt to the need to provide universal income in a jobless future they may help bring about a populist backlash against free trade and industrial modernization."

Those thinking that the planned end of work scenario is just a utopia should remember that in pre-industrial societies there were much less working hours than today, if for no other reason than people could work only during the daylight. Besides, they benefited from more religious holidays during the year. If now we work so hard, it is because of the invention of gas and electric lighting which has artificially extended the working day, especially in winter time. The introduction of machinery has done the rest. From the point of view of the capitalist, it makes no sense to buy expensive machines and turn them off at every religious celebration or just because the sun goes down. The prolonging of the working-day at the beginning of the industrial era was analysed in detail by Karl Marx (1976, 526–43). But this trend began again in the twentieth century. Indeed, sociologist Juliet B. Schor (1993) remarks that "one of capitalism's most durable myths is that it has reduced human toil. This myth is typically defended by a comparison of the modern forty-hour week with its seventy- or eighty-hour counterpart in the nineteenth century." The problem is that "working hours in the mid-nineteenth century constitute the most prodigious work effort in the entire history of humankind." In the preindustrial era the situation was much different. Just to give a few examples, a thirteenth-century estimate "finds that whole peasant families did not put in more than 150 days per year on their land. Manorial records from fourteenth-century England indicate an extremely short working year – 175 days – for servile laborers. Later evidence for farmer-miners, a group with control over their worktime, indicates they worked only 180 days a year."

There is no reason why a technologically advanced society should *force* its citizens to work harder than their ancestors, when they could work a lot less and without giving up their modern life standards. Among other things, this policy would also give workers more free time to take care of their children, the elderly and the disabled. Or they could just spend time with their families and friends, if the care of people is entrusted to robots. Unfortunately few are aware of the irrationality of our current situation.

According to anthropologist David Graeber (2013), the situation has now become even more paradoxical than before, because most of our jobs are not needed at all. It is worth remembering that "in the year 1930, John Maynard Keynes predicted that, by century's end, technology would have advanced sufficiently that countries like Great Britain or the United States would have achieved a 15-hour work week. There's every reason to believe he was right. In technological terms, we are quite capable of this. And yet it didn't happen. Instead, technology has been marshalled, if anything, to figure out ways to make us all work more. In order to achieve this, jobs have had to be created that are, effectively, pointless." Graeber surmises that this is not happening by chance. According to him "the ruling class has figured out that a happy and productive population with free time on their hands is a mortal danger (think of what started to happen when this even began to be approximated in the '60s)."

That being said, perhaps we should also consider possible negative collateral effects of the planned end of work scenario. Before reaching the final stage when stocks and dividend income are owned by citizens, societies are likely to attempt to redistribute employment by reducing working hours without a reduction in pay. Since firms would rather increase employees' working hours and reduce their pay, they are likely to react by threatening to relocate to other countries in our increasingly globalized economy. In its drive to pay workers as little as possible the private sector at the micro level is in conflict *with its own interests* on the macro level. It is in the interest of every company to employ a minimum number of workers, pay them a minimum salary and have the highest productivity. But if every company had what it wanted, in a closed system, there would be no consumers and therefore the same companies could not sell what it produced. Capital mobility in the global market has allowed companies to escape this dilemma. But this too, with time, will become a closed system, with the difference that there will be no regulators. In the nation-states, governments have always resolved the contradictions between the micro-rationality of companies and the macro-rationality of the economic systems, mediating between companies and trade unions, and regulating the labour market. But the global economy has no government, so national systems will not have any other option than exiting the global economy.

Indeed it is not hard to foresee that – were we to arrive at the absurd situation of a form of technological progress that generates hunger instead of wealth – nations would one by one withdraw from the global market to try to preserve their levels of employment. It is true that imposing reduced employee working hours on companies, while maintaining salaries at the same level, might convince them to move production elsewhere. However, in a future in which they employ almost exclusively machines and not humans, it would not be possible for companies to blackmail government and citizens by threatening to fire thousands of workers and they would have to transfer to more turbulent countries with chronic unemployment and rampant crime. These companies would therefore stand to lose.

Thus, if a certain degree of autarchy would weaken a degrowthist country, it would not be the same for a technologically advanced country with sufficient sources of energy and endogenous factors of technological development (brains and scientific institutions that are up to standards). A hyper-technological semi-autarchic state could maintain domestic order via the distribution of profits and the concurrent reduction of working hours, positing as an asymptotic ideal a society in which unconscious machines would work for sentient beings,[27] while sentient beings would devote themselves to recreational or higher activities, such as scientific research and artistic production.

A second unknown is the reaction of people who still work – let us call them the 'irreplaceable' – when seeing a mass of people paid to enjoy themselves and to consume. Moravec is clearly enthusiastic about the robots that he is designing and convinced that they will be able to do *any* work. But – if we assume that machines will be very sophisticated but still not conscious – its seems more plausible to think that, even if *any* work could in itself be executed by an intelligent robot (surgical interventions, repairs, artisanal works, transport of goods and people, etc.) there must always be a sentient being in the loop who acts as supervisor. Someone will have to be there, if only to act as maintenance manager of the machine, or of the machine that maintains the machine, or to gather data on the behaviour of the machine ('spying' to prevent unpredictable collateral effects). When trains, planes, and taxis are autonomous their users – for psychological reasons – will want to think that a human being is still somewhere in control.[28] However few these workers may be, why should they work when others do not have to?

The shrinking working class will likely be able to adapt to coexistence with the unemployed and accept that everybody has the right to live off a capital. Irreplaceable workers could receive a salary in addition to the citizens' wage, which would give them a higher social status in exchange for their job. Or, to preserve a sense of community, societies could institute a compulsory civil or military service that

---

27 A category that may include humans, transhumans, posthumans, machine-human hybrids or cyborgs, machine-animal hybrids, some animals, etc.
28 At least for the first generation of users, since later generations may have more trust in the machines.

would employ all citizens for a few hours a week to carry out functions of control and supervision.

Besides, it is highly probable that sooner or later all nations will be constrained to remedy technological unemployment in the same way, and, faced with increasingly homogeneous global conditions, borders could once again be opened – to promote at last the free circulation of people and goods. As communications and transports continue to develop, the world becomes ever smaller and borders ever more anachronistic. Therefore autarchy could be just a painful but necessary phase to overcome the resistance of capital to global regulation. And at the end of the process we shall have humans truly liberated from obligatory work.

3.7 An ethical judgement

We have seen that the social questions arising from the increasing robotization of industry and from the spread of artificial intelligence in the social fabric are similar in some ways to those that came with the mechanization of manufacturing during the industrial revolution, and are partly entirely novel ones. In both cases the process arises, mostly, within the context of a capitalist economy, and therefore the effects differ for different social strata.

Put more simply, robotization is not good or bad in itself but is good for some social classes or categories and bad for others, in accordance with the concrete effects that it has on people's lives. In a hypothetical perfectly equal society however the effects of robotization would be generally positive. Computers and robots can replace humans for repetitive or dangerous tasks. In addition they would allow a faster and cheaper production of consumer goods, to the advantage of the consumer. Thus the quality of the product increases, because of the greater precision of machines. Besides, some tasks can *only* be done by robots and computers. For instance, some work requires such precision that no human could do it (fabricating objects at the nanoscale for instance) and others can only be conceived and made with 3D printers and computers.

However, since societies today are highly layered, and some – for instance the Italian one – have a rather rigid class structure offering few possibilities to rise or fall, one cannot disregard the negative effects that robotization would have on some social classes without radical political and economic change.

To remedy this situation – as it should already be clear – our preference goes to the planned end of work scenario, for this is the only strategy that takes seriously the need to ensure a widespread distribution of the benefits of automation. The planned end of work scenario would also be the most *fair*. As I wrote in the journal *Mondoperaio* a few years ago:

Machines will be able to do most jobs, including designer jobs, even when they are deprived of consciousness or emotion (a possibility however that cannot at all be excluded). If no company will find it convenient to employ human beings, because they can be replaced by robots that work intelligently with no break, costing only the energy to run them, one will have to think of another social structure that could imply the abolition of work. Citizens could obtain an existence wage (or a citizen's wage) and be paid to consume rather than to produce. This solution would be ethically justified, because science and technology are collective products, that owe their existence to the joint effort of many minds, working in different places and historic times (Campa 2006).

This is the concept of epistemic communalism that I examined in-depth in my book *Etica della scienza pura* (Campa 2007). So I continued: "A quantum computer for example produced by a Japanese company would not have been conceivable without the ideas of Democritus, Galileo, Leibniz and other thinkers. In addition scientific research is often financed with public money. It would be unfair to tax workers in order to finance research the final result of which would be their own social marginalization."

In brief, the collective character of technoscience amply justifies a politics based on solidarity.

3.8 Conclusions

To sum up, we have outlined four possible scenarios. Two of them imply the end of robots, two of them the end of work. Among the last two, one scenario is dystopian, the other utopian. In the worst case humanity would be reduced to slavery under a capitalist elite. In the best case humans would live to consume and to enjoy reciprocally, while robots do the hard and dirty work.

The utopian scenario in its turn has two possible faces: one social democratic (redistribution based on social policies supported by taxation) and one socialist-capitalist (redistribution of ownership of robotic industries to citizens). One way or the other the forecast is that the whole of humanity will retire having worked just a little or not at all.

As regards the utopian scenario, one can observe that Moravec seems to have an enormous faith in people's ability to impose its own reasons and its own interests via the tools of democracy. A future that is not necessarily so uniform appears more likely to me, considering that neither the past nor the present have a single face. In other words, an intermediate scenario between dystopia and utopia seems more probable, with variations from country to country, from people to people, depending on political awareness, the level of infrastructure, and the degree of democracy. Contrary to what many futurologists appear to postulate in their analyses, human societies will not have the same future.

4. Roboethicists and Automata

4.1 Roboethics: a discipline in statu nascendi

The development of robotics and of automation has inspired the creation of a new academic discipline: roboethics.[29] The necessity of this field of study is deeply felt (Gunkel 2012, 2). The specialists dealing with the moral dimensions of robotics have been named 'robo-ethicists' or 'roboethicists' (Ganapati 2009; Lin, Abney and Bekey 2012; Fitter and Nichols 2015).

Let us start then with a definition, referring to one contributed by Antonio Monopoli (2007, 13): "We could define as roboethics that part of ethics that concerns itself with problems relating to robots and to their interaction with man, animals, society, nature and the world in general." He immediately specifies that "for the sake of our study we could divide robots into many categories: anthropomorphic and non-anthropomorphic robots, robots more or less similar to animals, robots with varying degrees of processing power, robots with varying degrees of ability to interact with the external environment, robots with or without organic components."

This definition will therefore be our starting point, even though – given future developments of robotic machines – it will probably be necessary to work out the goals and methods of roboethics in greater detail.

When we consider the emergence of this field, we face a classical dilemma. In order to decide when roboethics was born, should we focus on the existence of substantive and significant contributions (meaning simply what has been written on the topic) or rather on the institutionalization of the field (the proposal of a name for the discipline, the clear delimitation of the field of study, its presence inside academic institutions, the emergence of specialized journals, the organization of national and international conferences)?

If one focuses on substance one can – as usual! – go back all the way to the Greeks and to Aristotle. As we mentioned in the previous chapter, already the Stagirite stargazed about the birth of automata and imagined a humanity that did

---

29 It would perhaps be more correct to speak of a sub-discipline or field of study, because conceptually roboethics is just one branch of ethics. We think that it is simpler to see it in this light, also at the institutional level, because the emergence of biological computers, cyborgs, animal-machine hybrids and robots with organic parts will soon turn some roboethical issues into bioethical ones. There are no doubts, however, that roboethics, bioethics, and technoethics are fields of ethics of growing importance.

not need slaves. Potentially therefore, machines that were capable of autonomous work, even though only imaginary, carried a positive ethical value.[30]

A couple of millennia later, self-moving machines become a reality. Observing the situation in 19th century England, Karl Marx (1976, 502) remarked that "a system of machinery, whether it is based simply on the cooperation of similar machines, as in weaving, or on a combination of different machines, as in spinning, constitutes in itself a vast automaton as soon as it is driven by a self-acting prime mover." However, commenting on Aristotle, the author of *Capital* stated also that Pagans "understood nothing of Political Economy and Christianity. They did not, for example, comprehend that machinery is the surest means of lengthening the working day" (Marx 1976, 532–3).

Indeed – as David Ricardo (2004) also recognized – when the first industrial automatons arrived, the unforeseen and undesired effect was the mass dismissal of the workers and the longer working hours for those who remained in employment (given that their tasks were less strenuous; they only needed to look after the machines).

This discussion intensified in the 20th century, as more sophisticated forms of automatons, industrial robots, and humanoids emerged, and became a major topic at the start of the 21st century. In a way this whole debate amounts to roboethics because it concerns the 'good' and 'bad' that comes with the emergence of automatons and robots in society or in some parts of it.

Consequently, we could tackle the question of roboethics from a wider historical and philosophical perspective, but we will take another path. We will pay closer attention to its form and to the present. There is likely more than one who alleges to have invented the discipline. One of these is Gianmarco Veruggio, who claims that he invented the name, and therefore also the thing it stands for, in 2002 (Veruggio 2007, 7). Indeed, he helped developing the field of roboethics by publishing several articles on the topic, together with Fiorella Operto (Veruggio and Operto 2006, 2008) and alone (Veruggio 2010). Recognition is due, but not so much for having coined the name, as for his effort to launch a debate at an international level. On the site of *Scuola di robotica* (2011), which also offers the

---

30 "For if each tool could perform its own task either at our bidding or anticipating it, and if – as they say of the statues made by Daedalus or the tripods of Hephaestus, of which the poet says, 'self-moved they enter the assembly of the gods' – shuttles shuttled to and fro of their own accord, and pluckers played lyres, then master-craftsmen would have no need of assistants nor masters any need of slaves" (Aristotle 1995, 5). Also cited in Karl Marx's *Capital*, Book I, Section IV, "The Production of Relative Surplus-Value", Chapter 15, "Machinery and Large-Scale Industry" (1976, 532).

first bibliography in Italian, we read that "The term and concept of Roboethics was coined by the roboticist Gianmarco Veruggio who, in 2004, at Villa Nobel, Sanremo, organized a meeting between roboticists and scholars from all over the world, to discuss the theme of the ethics that the design and use of intelligent machines, or robots, generate."

Indeed a first attempt to compare ideas and perspectives in the field of roboethics has already taken place. On January 30th and 31st 2004, researchers in robotics, scientists, but also philosophers, lawyers, sociologists, anthropologists and moralists from all over the world agreed to meet in Sanremo, in the historical villa that once belonged to Alfred Nobel, and discuss the problems inherent in the relation of man and robot.

The First International Symposium on Roboethics was organized by the School of Robotics of the CNR in Genoa, presided over by Professor Veruggio and lead by Dr Fiorella Operto. Operto (2004) describes the event as follows:

> The goal of the conference is to discuss the preliminary guidelines for managing the man-automaton relation ethically and without causing harm, taking into account that world robot population is increasing exponentially: in 2004 there will be more than 970,000 spread all over the planet. There are many experts of the highest competence: there is the 'pope' of Japanese humanoids, Hirochika Inoue, Professor in the Department of the Mechano-Informatics of the School of Information Science and Technology of the University of Tokyo. Together with him are other well-known researchers of the Land of the Rising Sun: Kazuo Tanie, Atsuo Takanisi (both involved also in the Humanoid Project), and Tamaki Ura, whose field is marine robotics. Ronald Arkin, Director of the Mobile Robot Laboratory of Georgia Institute of Technology in Atlanta (USA), one of the big names in international robotics, launched the debate on the topic of smart bombs and of the humanoid robots that might replace them in coming wars.

This initiative was 'blessed' by no other than the President of the Republic Carlo Azeglio Ciampi: "Initiatives like this contribute to strengthen the new humanism founded on the principles of compatible development and of respect for the human person."[31]

The invitation also extended to intellectuals coming from the world of literature. Indeed Bruce Sterling, the guru of cyberpunk science fiction, discussed roboethics. This is all testimony to the very open minds of the organizers. Also there were Brian Duffy, robotic engineer at the Mit-Media Lab Europe; Dario Floreano from the Lausanne School of Polytechnics; and Paolo Dario, the director, of the

---

31 Given that the President spoke of a 'new humanism' and not just of humanism, it would be interesting to understand if this might also include some element of 'post-humanism' or 'trans-humanism' (Ranisch and Sorgner 2014; Campa 2010b).

Arts-Lab of the Scuola Superiore Sant'Anna in Pisa and head of a major project of Italo-Japanese collaboration on humanoids. The ethic visions vary with cultural origins, academic background and individual leanings, but everybody agreed on one point: if 20th century was the century of the machine, the 21st century will be the century of the robot.

Thus it would seem that Italy and Japan are rather sensitive to problems of roboethics. Indeed, in Italy initiatives and websites on this topic are ever more numerous, while Japan is working on ad hoc legislation, and therefore they already entered the advanced phase of 'robopolitics.' However, Italian engineers have no inferiority complex. Veruggio stresses that "Italian output in this field is among the best in the world and we have exported robots also to Japan. It is our chief asset, but we cannot do more than that" (Operto 2004).

4.2 A discipline concerned with futurabilia

Given that these are early days for roboethics, some have raised the problem of the point of this philosophical sub-discipline. Veruggio remarks that informing, discussing and working out the ethical rules is currently underway, with contributions by the same scientists who design the machines.

There is a risk that, in the future, fanatic Luddites, taking advantage of an ill-informed and frightened public opinion, might block all research and manufacturing of the kind. Thus he says: "The importance, and the urgency, of a Roboethics are to be found in the lessons of our recent history. Two of the most advanced fields in Science and Technology, Nuclear Physics and Genetic Engineering, have been forced to face the consequences of the applications of their research under the weight of dramatic and complex events. In many countries public opinion, shocked by some of these effects, have requested that work in both these fields be halted, or rigidly regulated." Consequently, "we should already now mind the future of robotics so that in the future they cannot accuse us of irresponsibility… we are not able today to put [Asimov's] three laws inside the brain of a robot. We are construing more capable machines, but we cannot guarantee that they will not be used to do harm. And the risk exists that in the future the discussion on this topic be manipulated, and politically exploited, by extremists and fanatics" (Ibid.).

Indeed, the fears expressed by Veruggio do not appear unfounded in the least, if one takes a look at recent Italian history. In Italy, techno-sceptics, through mobilising much of public opinion, have managed to block research and applications, first in the field of nuclear physics, with a referendum carried by the wave of emotions stirred by the Chernobyl accident, and then in some of the sectors of biomedicine, including in vitro fertilization and the research on embryonic stem cells, with the 40/2004 law. Not to mention the ban on GMOs and other biotechnologies.

Bennato (2004) expresses the same idea by means of an analogy: roboethics is to ethics what Formula 1 is to the automobile industry. "Why talk about roboethics while the robots are still sophisticated but limited computers? I think that the principle of the Ferrari is valid for the futurist part of technoethics, that is, roboethics. Meaning: why spend such a huge amount of money on Formula 1 races? Does it make economic sense? Yes, it does, and we all know it. Car racings are big open-air laboratories to test the technologies that will later be part of serial produced cars. The same applies for roboethics: the danger is not immediate, but it is desirable to familiarize oneself with these different modes of reasoning for when the 'robotic revolution' will take place."

Discordant voices exist however. Some say that it is as useless as it is premature to talk about roboethics at the present. The problem is that robots and nanobots are different from all preceding technologies. Nuclear power can be used to heat houses or to destroy them. The choice belongs to humans and hence it is they that the moral discussion concerns. But if the technological object becomes a subject able to make decisions, then it no longer makes sense to give the designer ethical recommendations. Thus Bonaccorso (2004) says: "Ethics is a result of consciousness and not the opposite, so if one wants to apply a discussion of the kind to robotic systems one must first accept that no engineer should ever expect one action rather than another; at most he will be able to try to correct errors, but it is the machine that will have to self-assimilate the new rules once it has filtrated them and adapted them to its own internal representation of its environment."

Bonaccorso stresses that this does not mean that a good advanced robot *ought* not to be built as a friend to man and that its implicit 'mission' *ought* not to be peaceful coexistence. The problem is that if the goal is reached, this means that the product is not yet an ethical subject, but a very sophisticated object that is still entirely controlled by humans.

> It is very important to keep in mind that when the goal of scientific research is subjected to any kind of ethical imperatives, then this goal cannot be that of bringing about novel creatures; at the most it can hope to meliorate the already pretty widespread automatons, exactly as happens in the field of automobiles or of telecommunications. Is there any point here to discuss roboethics? I don't think so. Let us wait for science to move ahead and, if ever one day we bump into a 'Terminator,' first we'll run away, and then, our minds at rest, let us discuss the problem and try to define all the rules that 'the new sons of man' will have to learn to respect!

Bonaccorso's argument is certainly interesting and – within a certain conceptual framework – it makes sense. However it is at odds with our point of view (and also that of Veruggio, Bennato, Monopoli, and Operto), be it only because his conception of roboethics and of robots is different. One must understand that, for Bonaccorso, a robot raises ethical problems only if it is conscious and fully autonomous. But the discourse changes if the robot is thought of in more general terms, and if one accepts that the receivers of ethical recommendations are primarily designers, construers, owners, and users. As Veruggio (2007, 7) put it, "Roboethics is not ethics for robots, but rather for robot experts and robot manufacturers."

Hence an ethical question, while raised in different terms, remains in the field as to whether the robot ends up entirely controlled by humans, or is partly autonomous and conscious, or appears as such even when it is not. In addition, the problem of autonomy and of consciousness can also be formulated 'by degrees' and not just in 'discrete' yes-or-no terms. It can also happen that a human being may find himself in an intermediary state between full awareness and no awareness. One may think of certain comatose states, of sleep-walking, of altered states of consciousness due to the intake of alcohol or drugs, and also of the various states of consciousness of a child who has not yet mastered the language of a fully grown adult, of an elderly person suffering from a degenerative illness, or of a person suffering from mental illness, etc.

Therefore, one may think of roboethics in different terms, and include not just the prescriptive aspect but also the descriptive one, making both human behaviour towards the robots and that of robots towards humans the object of its study, and conceiving of 'consciousness' in gradual and not metaphysical terms. In each case, it is appropriate to speak of behaviours and not of actions (a concept that implies a will), precisely because the category of robots is not limited to those endowed with a human-type consciousness.

In the end, one must avoid conceiving of roboethics as a discipline the purpose of which is to *do away with all possible risks.* If we put the problem in these terms then discussing roboethics becomes pretty pointless: since it is impossible to eliminate all risks, we have lost right from the start. But if the goal is to develop these new technologies *while trying as much as possible to prevent unnecessary risks,* albeit in the awareness that it can fail, this means that no stone must be left unturned and that the ethical debate is as useful as ever.

Robots exist already and they are interacting with human beings. For decades industrial robots have been working side by side with blue collar workers. Non-humanoid robots are already used at the workplace, on the battlefield, in spatial missions, to take over tasks that are risky or impossible for humans. And the first humanoid robots that take care of the elderly and the disabled, and help them face the challenges of daily life, are becoming available. These machines are not conscious in the same way as humans are conscious, but we think that studying their behavioural interactions with humans is extremely interesting both ethically and scientifically. We are convinced that, if and when the first Terminator will appear, it will be easier to counteract if there already exists a network of experts who for years will have been working on the problem and if, in the meantime, public opinion will have been adequately prepared to face this possibility.

If one understands roboethics as a discipline of futurabilia, which concerns itself not just with present problems but also possible ones in a hypothetical future, then we think it would be advisable to redefine it in wider terms. Generally one assumes that roboethics consists in defining some of the rules that must limit *the behaviour of the robot* so that it does not harm humans. One example is the famous laws of robotics worked out by Isaac Asimov (that we will look at in detail). Inside this perspective, the ethical problem is directly transferred from the acting robot to its maker. Rather than the machine itself it is the maker and the user who are required to assume 'ethical' behaviour (or 'ethically acceptable'). Ultimately therefore roboethics regulates *makers' behaviour*, to prevent them from deliberately or recklessly creating a hostile artificial intelligence. This however is rather obvious.

However, we should wish to stress the fact that the hypothetical creation of beings endowed with autonomous action generates more serious ethical problems. One aspect of roboethics appears to be overlooked in many debates. There are at least two schools of thought: one holds it that robots will always be machines deprived of consciousness, basically sophisticated electro-servants; another argues that sooner or later we shall be able to create machines that are sentient, conscious, or – even if they are not – able to behave *as if they were.*

Roboethics reply is different in these two hypothetical situations. If we build an electro-servant, we can have absolute power over it. If, instead, we build a sentient robot we will not be able to do what we want with it. We will also have to introduce rules regarding *our own behaviour towards machines.* These rules will have to constrain not just the behaviour of makers, but of all humans interacting with the robots. In other words, creating a conscious machine includes in itself a self-limitation. By analogy, when creating humans and giving them free will, Prometheus or the Biblical God accept ipso facto to limit the possibility of their own influence on the course of events.

Here, we would want to enrich the discussion on the definition of roboethics by proposing the following formula (not necessarily an alternative to the one mentioned above): roboethics is that branch of ethics that studies in descriptive terms the behavioural interactions between human beings and sentient and nonsentient, humanoid and non humanoid robots, and which normatively regulates the voluntary actions that humans can exert when dealing with robotic machines, including the act of its creation, and that the machine exerts in its interaction with its creator.

4.3 Roboethical codes

Once the necessity for a reflexion on roboethics has been recognized, we need to establish the ethical code that is to guide us in making decisions, stipulating conventions, formulating advice. It would indeed be naïve to believe that what is ethical and what is not is self-evident. Were it so, there would be no necessity whatsoever to open a debate. Makers and users of robots would already know what to do and would already be working for the sake of the good. Or, on the contrary, aware of the unethical nature of their undertaking, they would be at work underground. This is not the case.

Unfortunately, when we read newspaper articles on the new technologies, we often come across this naïve attitude. We do not infrequently stumble upon a journalist or a politician who states that: "This is not ethical," without feeling the slightest necessity to define even summarily his/her own ethical perspective or give some specific reason. A simple observation would suffice however to make clear that every ethical judgment requires a rational justification: if a large number of people behave in a certain way (for example they build robots armed for war), and they do this in plain daylight, then it is obvious that this group at least finds such behaviour ethically acceptable. In brief, it has its own reasons. These do not necessarily have to be accepted, but every criticism must rest on explicit and rational criteria.

There is no time to venture into a discussion about the fundamental theoretical problems of ethics, or to give our overall vision on morals. It would go beyond the scope of this inquiry. The minimum rule we lay down here is that we will take into consideration only those ethical propositions (ours and others) that are founded rationally – in the sense that they are supported by explicit rational and empirical arguments – and, above all, oriented towards solving specific practical problems. In this sense, we can say that our approach is rational and pragmatic.

In addition, we will refer to a 'classical' notion of ethics, that is, one in line with traditional Greek philosophy, which does not limit ethical behaviour to just altruism. Of course disinterested philanthropic behaviour constitutes a privileged form of ethics, but prudential norms, that is, aimed at accomplishing good for the individual actor or for the actor's community, are also part of ethical discourse.

In more general terms, ethics is the knowledge and practice of 'good' (which for Aristotle boils down to the concept of 'happiness'). Therefore, the ethical act is ultimately founded on benevolence (the will to do good), but this act of benevolence can target a variety of subjects: oneself, one's family, one's community (city, party, class, nation, empire, etc.), all mankind, all sentient beings (higher animals, hypothetical aliens, highly sophisticated machines, etc.), the set of living beings (which includes vegetables, insects, invertebrates, etc.), or even all existing beings (including mountains, lakes, rivers, planets, stars, etc.). When one takes into consideration a form of benevolence that also concerns itself with non human beings, then we can speak of a trans-human ethics. Ethical dilemmas arise when there is a conflict, that is, when a benevolent act toward someone ipso facto becomes a malevolent act toward someone else. Unfortunately, given the complexity of the world, these situations are more frequent than one would wish. Only die-hard idealists ignore this. Mostly ethical discords arise from the ambivalence of much of our behaviour, precisely on matters of benevolence and malevolence.

Since we are aware of this, we give up any hope of convincing everybody right from the start, even more so because we are tackling two topics that are already heavily loaded ideologically: technology and war. Part of the population holds ideologies that are a priori incompatible with the idea of setting up a rational ethical framework that can regulate the use of the new technologies in the context of an armed conflict. We are referring to the 'Luddism' that regards as an absolute evil the entire spectrum of technologies springing from the industrial revolution (and sometimes also from the Neolithic era) and the 'radical pacifism' that regards any war as an absolute evil, even if undertaken in defence. For these two groups, the locutions 'ethics of technology' and 'ethics of war' are two oxymora, two contradictions in terms. We will not discuss the validity of these two points of view. This has been done in other writings of ours. Our apologies to luddists and radical pacifists, but it is not our goal here to discuss general problems such as the good or evil of technology overall, or even to decide whether or not there are such things as just wars. Our goal is much more modest.

We take note that there are technologies and there are wars. And that there probably will be also in the future. These two phenomena generate ethical dilemmas. We will limit ourselves to the formulation of well-argued judgements on the dilemmas that arise, in the hope that our opinions may convince, if not everybody, at *least a majority of interlocutors*. In this sense we also accept the relativity (or the non universality) of our point of view.

As noted above, a discussion aiming to elaborate a normative ethical framework, or even a legal one, to guide the development of robotics is already systematically underway, involving academics and politicians of various leanings. In addition, there are many signs of good will from the makers themselves, possibly because they worry that the public opinion will react irrationally to the whole process.

The detailed reconstruction of all the propositions that have been advanced, every position in the camp, and the statistical magnitudes of every individual standpoint would be instructive. Unfortunately because of shortage of space we will have to limit our analysis to a qualitative one that takes into consideration only the positions most recurrent in the media and literature, or that we personally find most interesting.

4.3.1 Asimov's three laws of robotics

The most famous set of norms of roboethics – and at the same time one of the first to appear – is the one Isaac Asimov puts forward in his "Three laws of robotics":

1) A robot may not injure a human being or, through inaction, allow a human being to come to harm.
2) A robot must obey the orders given to it by human beings, except where such orders would conflict with the First Law.
3) A robot must protect its own existence as long as such protection does not conflict with the First or Second Law.

Asimov's three laws first appear in the short story "Runaround," and were then included in the anthology *I, Robot*, which in 2004 inspired a film with the same name by Alex Proyas, featuring Will Smith. In the Russian writer's imaginary world the Three Laws are deliberately codified into the positronic brain of the robot. Asimov calls 'Asenion robots' the class of robots that follow the Three Laws. This classification implies that there may exist other robots able behave in ways that disagree with the code.

In addition Asimov masterly shows all the paradoxes, the misunderstandings, the ambiguities that may arise from the interaction between robots and humans, in the presence of these laws. In one short story, for example, he shows the undesirable consequences that can follow from the First Law: a robot cannot assume the function of a surgeon because he would then harm a human. Similarly he cannot design strategies for American football because these might cause injuries to the players. To avoid this and other problems, the laws are not regarded as essential to more sophisticated robots.

Giuseppe Bonaccorso (2004) remarks that

> the literal execution of these laws is very often at odds with human morals: imagine that a robot takes part in a fight between two people and that at one point one of these takes out a revolver and threatens to kill the other. What should the robot do? Apparently it should intervene to save the life of the unarmed man, but this does not guarantee that its intention will succeed: both may become the victims of the bad guy who, feeling

threatened, would have to shoot without even thinking about the consequences. Surely a good negotiator would act differently... No program is up to evaluating all the possible hypotheses in real time, and only empathic consciousness (insofar as it is able to rule out a priori any ridiculously improper option) is apt to explain to a hypothetical bystander, be it human or artificial, that a few well-chosen words are more than enough to disarm the man with the revolver.

Bonaccorso's example is particularly important to our argument. But we would also like you to take into account the possibility that only killing will stop the bad guy. And that, precisely because of Asimov's First Law, the robot can only opt for the path of negotiation. That is, the wrong one. More specifically, on the battlefield the robot is called to defend its own human co-fighter against the threat of other humans and negotiation is not an option. If it cannot harm humans (even enemies) how could it fight? Asimov's three laws appear made for a world at peace rather than a world at war. In any event, they prohibit using robots in warfare.

Let's consider an even more extreme case: if a group of terrorists (like Aum Shinrikio) suggested the eradication of all humanity, what should the robot do? Should it remain inert in order also then not to harm human beings (the terrorists)? This is the kind of ethically ambiguous situation that we mentioned above: benevolence towards one subject often implies malevolence towards another subject.

Perhaps, and just to tackle this kind of problem, Asimov extends the list of rules by introducing an even more fundamental law, which he calls 'Zero Law,' because it takes precedence over the laws with a higher number:

0. A robot may not harm humanity, or, by inaction, allow humanity to come to harm.

Here an abstract concept is introduced which forces the robot to think in more universal terms. This law is mentioned in the novel *Robots and Empire*. However it does not at all solve the problem. Giskard, the robot, in order to obey this law, effectively violates the First Law, which leads to the destruction of its positronic brain. The incompatibility of these principles, when applied to reality, make us see the importance for the development of roboethics of much of Asimov's literary work and of those writers of science fiction that have explored this topic.

We should keep in mind that the significance of Asimov's work goes beyond the limits of science fiction. Indeed the Japanese government imposes a series of rules to makers, inspired by those very laws formulated by the Russian author, in order to prevent harm to humans. Thus the daily *la Repubblica* reports:

Rather than worrying whether the robot be ever more similar to the splendid Rutger Hauer of Bladerunner, the Japanese government sees to it that it will not have his destructive powers. In a reality that is fast approaching science fiction, the Tokyo

Ministry of the Economy, Commerce and Industry is working on a list of security rules, which will be completed by the end of the year, and imposed to all companies building the robots of the next generation. One worries that ever more sophisticated machines might prove dangerous to humans, whence the necessity to establish criteria of security for their design It is truly a page of science fiction come true: in 1940 Isaac Asimov, the author of the Galaxy Trilogy and the writer who has best defined the relationship between man and robot, formulates the first law of robotics: "A robot may not injure a human being or, through inaction, allow a human being to come to harm." The guidelines being worked out by the Japanese ministry have precisely this goal, since they require the producers to fit the robot with an appropriate number of sensors that hinder them from crushing people and favour soft materials that reduce the impact upon collision. Another rule obliges them to endow the robots with on-off buttons that are easy to reach and use. The reason behind this prescription is understandable, if we think of all the science fiction books in which a poor human fails to stop a robot gone mad (Nadotti 2006).

The article then stresses the fact that, in Italy, where humanoid robots are rare, the Japanese initiative really appears as verging on science fiction. Where on the contrary the robot is already part of the human landscape, the introduction of this kind of legislation became a necessary response to citizens' requirements. In The Land of The Rising Sun, where there is a chronic shortage of assistants and nurses, the robotic industry is making very rapid progress in order to create automatons able to replace humans in the task of assisting the elderly and the sick.

In Japan the problem of caretakers could hardly be dealt with in the same way that it has in Italy, that is, via massive immigration from economically disadvantaged areas. A look at its demography and its geography will show that this Asian country is in fact relatively overpopulated. Over 127 million inhabitants live on an area this is barely larger than Italy, with a density of 337 inhabitants per square kilometre (versus 199,9 inhabitants/km^2 on our peninsula). And Italy is itself a densely populated region compared to other regions of the planet. It is therefore not astonishing that Japan prefers the humanoid to the immigrant and gives a great impetus to robotics industry.

The latest invention of Japanese technology are: Ri-man, the robot that can take care of the disabled, and Wabian-2, the robot that can vary its facial expressions with the state of mind it wants to express (the latter will be manufactured in Italy). In Japan there already exists a law regulating the production of industrial robots (Occupation Health and Safety Law), but one also feels the need for a specific legislation regarding the latest generation of house robots. Given that in Japan the total turnover of the robotics industry in 2005 was over six billion yen, the zeal to regulate the manufacturing of robots ethically and legally is hardly astonishing. Cristina Nadotti (2006) stresses that "in the end of 2004 there were more than 356,000 industrial robots in use in Japan, the highest number in the world. To give an idea of their advantage over other nations, it is enough to measure the distance that separates Tokyo from the US, the world's second robot-using nation, with 'only' 122,000 robots." The METI – an acronym standing for the Japanese Ministry of Technology – forecasts that by 2020 the turnover of the robotics industry will be higher than the car industry.

4.3.2 The Euron Codex

There are codes of roboethics other than those of Asimov, which – as we have seen – is hardly compatible with a belligerent use of robots and is at the root of many paradoxes. One of those codes had been has been worked out by an international group of scientists at the head of EURON (European Robotics Research Network) and which we therefore temporarily refer to as the Euron Codex. In the words of Gianmarco Veruggio there are two closely related priorities: "We must work out the ethics of the scientists who build the robots and the artificial ethics to build into the robots. The scientists must begin to analyse this kind of problems and decide if laws or regulations are necessary to protect citizens. Robots will evolve to become highly intelligent, so that, in some ways, they will be more intelligent than humans. But their intelligence will be an alien one. For this reason I would prefer to prioritize humans" (Habershon and Woods 2006). Therefore one feels it is necessary to tackle the problem at once, before robots become more intelligent, faster and stronger than man, as well as massively present in our society.

The strategy of giving priority to man is in harmony with that of Asimov, who always conceives of the robot as being in our service and under our control. However the Euron Codex is not yet working at the formulation of detailed laws and regulations, but focuses on articulating general recommendations, from which more specific regulations can later be derived. Among the priorities we find the following: manufacturers must ensure that machine actions remain under human control; one must prevent unlawful use of the machines; one must protect information data obtained using the robot; machines must be strictly identifiable, as well as traceable.

Here is the synoptic table of the five roboethics recommendations:

Safety	Ensure human control of robot
Security	Prevent wrong or illegal use
Privacy	Protect data held by robot
Traceability	Record robot's activity
Identifiability	Give unique ID to each robot

In other words, these are prudential norms that seek to predict, discourage or prevent possible immoral or illegal use of the machines, and even to react efficiently to their hypothetical 'going mad'. Mafias, terrorist groups, criminals could indeed use robots to kill, rob, intimidate, threaten and blackmail. A robot is not afraid of being arrested and punished, and hence, because it has fewer psychological and operational restrictions than a human killer, it is much more dangerous. This is why it would be useful to make an inventory and a catalogue of all the robots in circulation, determining the guardianship and the responsibility that befall their owners and to outlaw the buying or selling of robots that have not been licensed by the Register Office. The same as we do with means of transportation which, let us not forget it, are one of the main causes of death in modern society. A Register Office would not solve all problems, but it would help limiting them.

The Euron recommendations may be a good starting point, but – as we will see in what follows – fulfilling these fundamental principles runs into practical problems with no simple solution. First of all it would be necessary to clarify what is meant by "controlled by humans." In the popular imagination, a robot is not just a remote controlled machine, but a machine that more often than not has some degree of decision-making autonomy (even though confined within a program and specific environmental conditions). Control may therefore go from humans deciding every single action of the robot to situations in which the robot is basically 'free,' but can be switched off at any time by humans. Therefore, the concept of control is necessarily gradualistic. On this scale where is the ideal point to be situated?

In addition we need to ask ourselves: "controlled by humans" in what sense? By any human being? Or by the lawful owner of the robot? And if these are absent, who should exert control?

If we reply: "by any human being," then the possibility of using the machine for warfare, or for surveillance, or in any activity (also economical) that favours one human over another, is ruled out. If *all humanity* were empowered to control *all robots* why acquire the robot? It seems more reasonable to think that it will be the owners who will do as they please with their robots, switch them on and off, and using them (hopefully) without breaking the law. But it will never be possible to literally speak of humanity's control over robots. Indeed, those robots used for surveillance and security turn this paradigm on its head, given that it is in fact the robots that 'control' humans, for the benefit of other humans.

Every technology favours some groups of people (in general those who own it) and is unfavourable to other groups of people (those who do not), by virtue of Francis Bacon's law: *scientia potentia est*. Therefore it is pretty naïve to wonder if a technology, including that of robots, is good or evil for humanity as a whole.

In addition, as regards the problem of 'safety,' one should take into account that 'human control' of robots does not guarantee anything if one does not specify the moral characteristics and the finality of just that human who exerts control. Mafias and terrorist groups are also parts of 'humanity.'

To sum up, it is correct to speak of a control system, of security, of privacy, of traceability and of ID, in order to prevent morally improper use of robots, but one has to examine the matter very circumstantially, keeping in mind the issue's sociological complexity.

As for the details of 'wrong' or 'immoral' behaviours to foresee or to prevent, we would recommend remaining within the framework of a rational ethical approach – which means that we should never assume that something is obviously good or evil. When one intends to limit or prohibit a certain type of behaviour one must give solid reasons that prove the harm to society that would ensue. Effectively many ethical codes that we inherit from the past consist in sets of rational norms, that we understand and share to this day, and of irrational norms, that is, founded on mere habits or superstitions that no longer ensure, if ever they did, increased happiness, well-being and overall good. For instance, if we take a look at Sumerian ethics (a reference to the first civilization that has handed down written records), we discover that this requires citizens to assist orphans and widows, to respect contracts, not to deceive customers with false weights and measures, not to steal others' food and drink, etc. (Kramer 1997, 102–109). These norms still make sense today. They have a motive: they want to grant a minimum of happiness, well-being and good to the most unfortunate, considering that this aid would not be too heavy a burden to the luckier ones, and also to respect others' property. However, by analogy with what happens in other ancient codices, in Sumerian ethics there are also rules concerning 'forbidden food,' 'forbidden places,' 'forbidden speech' and 'forbidden acts' that we today find hard to understand. These appear founded on superstition or on error. The only fact that they are hard to understand, rationally and emotionally, does undermine their universal character.

Some may ask: what does this have to do with robots? We reply with a detailed example. Some inside the EURON group started a discussion about the erotic use of robots. Thus Henrik Christensen stated that: "Security, safety and sex are the three big concerns." While concerns regarding control and safety are easy to understand, it is less obvious why sex could be a problem. He added: "Within five years people are going to have sex with robots." This had *The Sunday Times* ask the question: "Should limits be set to the appearance, for example, of these robotic sex toys?" and gave an answer in the headline: "No sex please, robot, just clean the floor" (Habershon and Woods 2006). *Punto Informatico* (2006) took up the *Sunday Times* on this issue: "What happens if a robot becomes sexually attractive? What happens if people begin to acquire humanoid robots merely for carnal pleasure?".

Now, we do not know to what extent this is what the EURON roboethics committee, and the journalists who tried to interpret their vision, are really worried about; nevertheless a possible prohibition to build or use robots for sex would be hard to justify on the basis of rational ethics. One suspects that it is rather residues of ancient ethics, of a sexophobic nature, that play their part here.

People who acquire robots for sexual gratification will merely exert the right to reach a good (their own happiness), with no harm to other people. Not only classical ethics but also current legislation supports this interpretation. However bizarre the practice in question may seem, no law prohibits having sex with objects. Nor could it because such a legislation would violate human rights, in particular the combined articles 1 and 4 of the *Declaration of the rights of man and of the citizen*: "Men are born and remain free and equal in rights" and "Liberty consists in the freedom to do anything which injures no one else." Moreover this declaration specifies that "Law can only prohibit such actions as are hurtful to society." Therefore one must first demonstrate empirically that this type of behaviour represents an existential threat to society, and not a mere change in its customs. And, even if the threat is there, it is not given that prohibition represents the most efficient remedy or a legally possible course. To give an example: even if one were to observe a drastic demographic decline and even if one were to prove the correlation between a decline in number of births and the erotic use of robots, a government could opt to create incentives to procreation, rather than to prohibit non-procreating sexual behaviours.

Naturally one may legitimately regard robotic sex as a problem, but on condition that one makes it clear in what sense this is a 'major issue.' Indeed we may ask ourselves how one can reconcile an erotic use of the robot with privacy and traceability. If everything the robot does is recorded and available to some authority, its erotic usage would be strongly limited. On the other hand, its traceability is there to prevent a criminal usage. This is a genuine dilemma that is worth discussing.

The question regarding the part of robots in maintaining public order also offers a lot of food for ethical and prudential thought. Ronald Arkin of GalTech argues that it "is necessary to evaluate what might happen if robots were empowered to control crowds and authorized, for instance, to violently repress a mass uprising" (Ibid.).

In this case we would go beyond the ethical horizon of Asimov's Laws, but this use would still be legitimate given a certain interpretation of the Euron Codex (that is, "control by humans" would be understood as the capability *for the owner*, and not for just anyone, to stop the machine at any time). The question of the hypothetical repression of a rebellious crowd would therefore be evaluated in the light of the laws in vigour in a given country. The robots should not be permitted to do anything the police itself would not be permitted to do.

The main question however is that of responsibility. If a policeman acting to repress abuses his power and inflicts gratuitous violence on demonstrators, he can be singled out and punished. In the same way, if a policeman wounds or kills a demonstrator, but that it is proven that he acted in self-defence (to preserve his own life and safety), then he cannot be punished or he would have mitigating circumstances. But in the case of the robot there can be no such mitigating circumstances since it is neither alive nor conscious. Therefore, if a system of robotic weapons were used for public safety and wounded or even killed a human, then the necessity arises to establish someone's responsibility and someone's guilt. If the makers or sellers are held responsible, these will probably claim that using the robots to this end is improper and reject their responsibility. If the magistrate decides that it is the commander who sent the robot to repress the crowd who is responsible, the robot will probably no longer be used for such tasks. They will be used for surveillance, but physical contact with humans will be restricted to human policemen. At least in nations subject to the rule of law.

The ethical problem remains an open one for a similar use of robots in totalitarian regimes. And it might arise again in democracies whenever the nature of the robot or of artificial intelligence changes. The well-known futurologist Ian Pearson states that within a few years we will no longer be able to treat robots like we do objects: "My forecast is that we will have conscious machines before 2020." In five years that is. "If we endow robots with consciousness they will become androids. This will constitute a huge ethical change" (Habershon and Woods, 2006).

4.4 Evolution and legal responsibility

In nations where the robotic industry is fast expanding, governments have over and over laid down norms regarding the design, making and use of robots, taking their inspiration from codes worked out by experts. Nevertheless, the robotic legislation in force is often obsolete a few years after its formulation, proving that the belief that one can lay down the rules once and for all in this area is an illusion. Robots evolve. And they evolve faster than human beings.

Roboethicist Monopoli approaches the problem of robotic responsibility from an evolutionary angle. Today robots are regarded as little more than home electric appliances, and hence the responsibility for possible harm that they might cause befalls the designer, the maker, the seller or the user. But will this always be the case? "One aspect of particular ethical relevance concerns the responsibility (or lack thereof) of the individual robot, relative to its own actions and their consequences. It is likely that, with time, robots will acquire ever greater ability for self-learning, 'understanding' and interaction with the external world; in other words we will have robots able to 'decide' what to do in any situation it will find itself in. This is a condition it has in common with humans who often have to face novel situations" (Monopoli 2007, 39).

It would be hard and even unfair to lay the guilt on the designer if the action of the robot is due to the personality of the machine, which has been formed via a slow process of learning and interaction with people around it. In the same way that parents are today considered responsible for the misdeeds of their children *up to a point* (until they cease to be their guardians, until their children reach maturity), robots give rise to an analogous situation. The question then arises as of *when* the designer and owner are no longer responsible for the action of the robot.

Given that robots will be manufactured among other things to take care of the elderly, we can examine the case in which a robot drops, and thereby permanently harms, the person it is assisting. If the assistant is a human then we face two types of penal and civic responsibility: that of the assistant and that of the company he works for. If the carer is instead a robot the case becomes more complex and is likely to alter with time. Let us take a look at the possible scenarios Monopoli (2007, 39) outlines:

1) FIRST HYPOTHESIS: The robot is looked upon as a machine (for example a meat grinder) so the responsibility can only befall its owner and/or operator. The victim can accuse those who sold him the robot and the seller can in his turn blame the entire commercial chain all the way up to the designer, according to the norms in vigour in each nation.

2) SECOND HYPOTHESIS: The robots has extensive abilities for self-learning and interaction with the external world, and society now supports the idea of robots having some operative autonomy, because a given robot evolves following his own individual history of experiences; consequently harmful behaviour can be seen as a fortuitous and unforeseen event, with no direct correlation to other identical robots. In this case one could invoke the good faith of those who designed and commercialized the robot and view the event as unpredictable and accidental. However, to avoid the victim having to pay for the injury, it would be advisable to set up an insurance fund to cover cases like this. In any case it is clear that whoever will formally have to pay, it is the end user who, in some way or other, will have to bear the cost.

3) THIRD HYPOTHESIS: Robots have reached a level of 'intelligence' in situations and interactions with the external world such that it is able to make decisions in complex

situations; in addition they have 'an internal life of the mind' able to autonomously generate choice criteria. This case can be regarded as a stage following from the former, and perhaps then we can, with much caution, speak of a situation of 'individual robot responsibility' the consequences of which could be treated in a similar way, on some points, as that of the liability of a company. In such conditions each individual robot should be regularly inspected in order to prevent deviant behaviour or the robot 'going insane.'

The hypothesis of a civil or penal procedure launched against robots may seem absurd, but we must keep in mind that the robots of the future may possess a juridical personality and also an ontology not unlike our own. They will not necessarily be machines of metal, plastic and silicon. Organic parts may be grafted onto the robot to improve its performance. The scientist Steve Potter has successfully implanted neuronal cells extracted from a mouse embryo in a non humanoid robot and obtained a machine capable of learning from its own mistakes and to plan creatively. The press has called it a robot with a mouse brain, although this definition is improper (Rampini 2003).

One can also envisage the situation in which the biological brain in question is not animal but human. Even if everybody accepted the ethical rules that it is prohibited to create computers and robots with human brains, this might still be achieved in various ways with the consent of the donor himself. As Monopoli (2005) stresses, "in our culture the concept of individuality is ever more often identified with our brain; simply stated, nowadays we are all lead to think that if we replace a hip with a prosthesis, or even if we add a prosthesis to the breast or an artificial hand, we remain the same always and anyway and this as long as our brain is able to function. The hypothesis then that in the case of disease or devastating accidents for the body in which the head or even just the brain could be 'saved' if there could be an intervention replacing the body, in parts or in its entirety, with an artificial one is therefore not so farfetched."

In this case we would have a robot with a human brain, rather than with a silicon brain which is the currently the case, or positronic as is the case in science fiction. In this situation it might seem unethical to prevent, rather than encourage this kind of technical intervention. Monopoli (2005) agrees: "Ethically this kind of intervention would paradoxically not present any particular problems since it respects the general principles justifying a therapeutic intervention, which is that one is faced with a genuinely pathological situation, that the therapy benefits the subject it is being applied to, that the kind of intervention is proportional to the gravity of the pathology, and there are no less invasive technologies that would offer similar prospects of a cure."

In truth, Italian law explicitly prohibits brain transplants, but the problem could be circumvented by arguing that it is a body and not a brain transplant. This shows how shortsighted and philosophically naïve Italian law is in matters of transplants. The inadequacy of the law will reveal itself even more when mind transplants, and not that of the mere brain, become a reality, that is, the mind-uploading that some scientists and experts in artificial intelligence dream about (certainly closer to science fiction, but not for this reason in contradiction with the materialistic metaphysical hypothesis and with our scientific knowledge). In this framework, the ethical limits on therapeutic interventions may become insufficient. If this kind of intervention reveals itself enhancing (aesthetically and physiologically) and gives the concerned subject entire satisfaction, also in the absence of any pathology, on what grounds could he be denied it? Here the idea of *one* unifying and shared ethics would vanish, and, as in bioethics, two or more ethics would present themselves. Someone would approve the meliorating enhancing treatment, because the interested subject would have freely opted to have it, while another might consider prohibiting it in the name of values beyond the individual: religion, race, humanity, nature.

Monopoli (2005) also takes into account the possibility of enhancing and not merely therapeutic interventions when he evokes the spectre of 'eugenic risk': "One ethical question would be that of someone with a malformed body wanting to replace it with a 'normal' one, and another ethical question would be that of desiring a new body simply because it is perceived as better or as more beautiful. Just think, for example, of the recurring idea of the 'super-soldier' or of the 'perfectly' beautiful woman who does not age. Obviously we are gradually slipping from the idea of therapy to that of a new eugenicism mediated by technology and it is necessary to recall that everything that is technically possible is not always ethically acceptable."

While we agree with Monopoli on many questions, we differ from him on this point. His analyses' shortcomings stem from the fact that he takes for granted that eugenics is immoral, and does not even feel the need to argue against it. The absence of a general argument happens when one presumes the existence of a self-evident, universal and just ethics (a very common situation among those who adhere to the perspective of a revealed religion such as Christianity, Judaism or Islam), or when one is unaware of alternative values and normative propositions that render this matter controversial. This by the way is understandable, given that one cannot have read everything. In any case, there is an extensive literature that argues in favour of liberal and positive eugenics. In addition, we must keep in mind that eugenics has an important cultural legacy in the West, as proven by works like Plato's *Republic*, Francis Bacon's *New Atlantis* or Thomas Campanella's

City of the Sun.[32] Of course today no one is suggesting that we argue in favour of the authoritarian and negative eugenics of the first half of the 20th century – which, for the record, was not supported only by Fascist States, but also by democratic and communist ones. The 'new eugenics' respects the person and above all the individual *will* and for this reason it calls itself liberal.

Liberal eugenics also faces opposition, including some secular prelates and intellectuals like Francis Fukuyama or Jürgen Habermas, but today there are many intellectuals and citizens who view it as perfectly legitimate. The rhetoric question that Gianni Riotta (2005) asks Francis Fukuyama (and the readers of *Corriere della Sera*) is significant in this sense: "Where is the harm in being born taller, or faster, the explosive cells of sprinting are innate, why not give them to everybody? And how could a better memory ever change our ethical code, or more widespread intelligence pose a risk to Homo Sapiens? I do not think that everybody would end up choosing Arian blond hair or glaucous eyes, the models of beauty are many and different."

Supporters of enhancing interventions, among these genetic mutations or grafting robotic prosthetics, always stress the necessity of free choice and free access to the new technologies, so that they do not become an obligation or the privilege of an elite. The objection that in reality the *free choice* is an illusion, because once some humans get enhanced they force everybody else to do so too, is very weak both empirically and rationally. First of all it is not true empirically. The Amish, in the United States, have decided to stick to 19th century technology, refusing electricity and internal combustion engines, and nobody forces them to go along with the rest of the world, or harasses them in any way. Secondly, the argument is also rationally weak because studying and sports improve the mind and the body of an individual, making him smarter and stronger. If it were evil to improve and to distinguish oneself with respect to others, then one could in all consistency prohibit the use of libraries and fitness clubs. To avoid inequalities, that anyhow exist from the moment of birth and are subsequently amplified by the willpower of some persons, then we would need to revise human society as a whole – and not just the capitalist one, given that intellectual

---

32 To understand that eugenics represents a controversial idea still today, and not simply one that is ethical or not, we will quote the piece by Peter Sloterdijk, "Rules for the human park" (2013). To give an idea of the current relevance of the ethical polarity in matters of eugenics, it is enough to confront it to a 2003 writing by Jürgen Habermas, *The Future of Human Nature*.

and sports competitions also strongly characterized historical Greco-Roman society and 20th century Soviet communism.

Roboethics will become highly controversial, as did bioethics, if one will want to apply the ethical rules of one social or religious group to all other groups, beyond any kind of logically and empirically founded reasoning. In a pluralistic, multi-ethnic, multi-political, multi-religious world such as ours, one cannot simply say "not this because it is unethical" without providing universally convincing arguments. If we are looking for rules that can be shared and held in common, then we must first of all be open in our approach to others' visions, and therefore it is good to be aware of the partial nature of our own convictions. One can only reach a shared ethics – if ever – through negotiation; not through imposition. Monopoli seems aware of this when he argues that ethics cannot be reduced to the law of economic interests (a rather widespread ethical vision in our society). In the same way, let us add, ethics cannot be reduced to a revealed religion or to habits. It must have a rational foundation.

4.5 Possible ethical problems in android robotics

Regardless of its still uncertain institutional status, roboethics is faced with very interesting philosophical and speculative problems having to do with the possibility that (more or less) conscious androids may cause problems to their makers, owners, users or human interactors. *De facto* we are creating ever more sophisticated machines. Designers devise these machines in the hope that they will be able to think and act similarly to and better than human beings, and one sees constant progress in this sense. This process cannot be without consequences or effect of an ethical, psychological, social and political kind. Let us then take a detailed look at the kind of androids that are being explored in novels of science fiction or in philosophical discussions, and the possible desirable or undesirable effects, intended or collateral.

4.5.1 The NDR 114 Model

In *The Bicentennial Man*, a short story by Isaac Asimov, made into a film with the same title by Chris Columbus in 1999, the main character – Robin Williams, alias NDR 114 – is a robot that becomes conscious of its own existence and undertakes a long technical and legal journey to become thoroughly human, gaining also the human trait of mortality. The android's gaining consciousness is presented as an unforeseen collateral effect, that is, one not willed by the designers.

This possible effect concerns first of all ontology, but it has clear repercussions for ethics. If, when we make a laundry machine, we are certain that we are making a home electrical appliance (a thing, an object), when making an anthropomorphic robot we might in reality be creating a child, that is, a conscious being who sooner or later will claim her autonomy. For the time being this is a hypothesis, but more than one philosopher, scientist, engineer and writer regard this outcome as a plausible one. Just to give one example, Hilary Putnam raised this question in a 1964 article with the title "Robots: Machines or Artificially Created Life?" And more recently, Stevan Harnad (2003), when asked "can a machine can be conscious?" simply replied: "A 'machine' is any causal physical system, hence we are machines, hence machines can be conscious." This to say that the real question is another, namely: which kinds of machines can think?

Within this framework, an engineer could perceive his relation to the robot in terms of parenthood. Today, even though machines do not yet give clear signs of being conscious, their makers do not really treat them like objects. They give them names and interact with them *as if they were alive*. In Japan, where cultural tradition is not very prone to regard natural objects as inanimate, one tends to spontaneously view androids as 'living' beings, and deserving some respect. Such an emotional relation can only increase in intensity as the machine gets more complex, and as the signs that can be interpreted as indicating the presence of consciousness in the machine grow in quantity and quality. From the point of view of how humans feel about it, the fact that the machine be truly conscious is secondary somehow.

After all, not only do we not know if machines will gain consciousness, but we do not know either if we will ever be able to tell for certain. The simulation could be so realistic as to remove any doubt. And, after all, we are not even sure that other people are conscious. The problem emerged in a conference entitled "Towards a Scientific Basis for Consciousness," that took place in the University of Arizona in 1994. During the discussion, the neuroscientist Christof Koch asked the philosopher David Chalmers: "But how can I know if your subjective experiences are like mine? How can I know if you are conscious?" (Horgan 1997, 182). If we cannot know this in the case of beings like ourselves, if we can only assume it, then we will never know for certain in the case of computers and of robots. The science journalist John Horgan (1997, 182) adds that "one can certainly imagine a world of androids that resemble humans in every respect – except that they do not have a conscious experience of the world."

The appearance of a model NDR 114 could become a moral problem primarily to someone who has a religious faith. If humans produce a thinking, conscious, desiring and feeling mechanical person, then they usurp the role that religions attribute to the gods. It is Prometheus (or someone equivalent) who creates humans in the polytheistic religions of Indo-European origin; it is Yahweh, God or Allah who creates humans in the monotheistic religions of a Semitic origin. Within a religious framework, when humans design life they challenge the gods, they turn themselves into gods. The creator of a model NDR 114 is, as one often says, 'man playing god.' This could lead to feelings of guilt and fear. At the same time the feelings of parenthood would create a situation of psychological ambivalence and internal conflict. The parents have a privileged relation to the children regardless of the latter's evil deeds.

It is clear that the most widespread worry, in particular that propagated by science fiction, is that the situation might get out of hand as regards public order, but the religious citizen might also fear a divine reaction. Indeed, a hypothetical god might perceive these creative acts as ones of arrogance, disobedience and threat. In reaction (s)he might cut off her/his alliance with humans and respond to their challenge by viewing them as enemies and annihilating them. In fact, some have also defused the problem from a theological point of view, stressing – as has Frank Tipler – that these creative abilities are anyhow given to humans by God, and therefore they could be part of His plan. In this respect the Catholic and Protestant interpretations differ, the former insisting on man's free will (and therefore on the fact that he could do something that God does not like) and the latter on predestination, which makes it easier to identify building robots with a divine intention. From this point of view the robotic engineers, and by extension humans, would be nothing other than the Creator's tools. We are however happy to leave this question to the theologians. Our intention here is only to point out that there exists a discussion of the sort.

The emotional effects of the emergence of a model NDR 114 could stir also those without religion. In this case, the designer would not become the object of God's wrath or guilt-ridden because (s)he would have competed with supernatural beings, but (s)he would still feel parental towards the machine. This problem is of no small significance. A parental attitude might lead the designer to be much more lenient and prone to forgive the 'child's' trespasses than would other humans. This might give rise to ethical divergences.

In other words, also when facing situations of real danger, the 'creator' could still build machines that he would endow with specific empowering features, arguing that fitting it with too many security devices would be excessively punitive or disabling to her/his 'creature.' That is, (s)he could refuse, more or less consciously, to limit or punish her/his own mechanical offspring. The problem might be overcome if one has the machine inspected by a committee of experts (engineers, philosophers, sociologists, psychologists) who would work alongside the designers. After all, at school, teachers decide of the promotion or rejection of pupils, without taking their parents' opinion into account.

In addition, many think that consciousness in robots is anything but undesirable, and so it should not be taken for granted that philosophers or sociologists will prevent the engineer from generating strong artificial intelligence. To uncover the secret of consciousness, of how it is created and manipulated, would also mean to uncover the secret of how to preserve it and transfer it to another support. It would mean eluding death.

4.5.2 The Galatea Model

Another interesting problem is what kind of emotional relation might arise between humans and automata. The erotic or sentimental relation is a problem, which, as we have seen, has been raised also by Henrik Christensen of the EURON group. It is pretty obvious that it is a problem pertaining mostly to robots intended for home use, rather than for industrial or military use. Unless one takes into consideration the hypothesis that in the future one could also send troops of "sex robots" to serve the army or local populations. Or that particularly attractive robots could be made to serve in intelligence, because it would also be capable of recording and transmitting information of any kind in real time. Let us however consider the matter in general terms. In the article "Roboethics: an emblematic case of technoethics," sociologist Davide Bennato (2004) calls 'the Galatea Principle' the issue of the sentimental man-machine relation and the corresponding moral worries and controversies, drawing his inspiration from the Pygmalion myth.

> Soon the moment will come when robot and human coexistence will be more deeply felt. The precursory signs are all over the place: recently in Japan sales of AIBO, the Sony robotic dog, have boomed and are gradually disrupting the pet market. Always in Japan, the National Institute of Advanced Science and Technology has set up a program of Pet Therapy called the Paro Project which, instead of using flesh and blood animals, uses robotic seal cubs. What is interesting is that these technologies will become the tools that will alter our emotional life. The principle of Galatea is valid here: Venus transforms the ivory statue into a woman in answer to the prayer of Pygmalion, the sculptor in love with his creature. We can derive a maxim from the myth of Galatea and Pygmalion that will begin our reflection on this topic: a deep emotional relation is real when the subjects involved are prepared to consider it real. Think of the relation that might arise between a person and the robotic version of an inflatable doll: what kind of relationship could it be? If you are sceptical, you may want to have a look at high-tech inflatable doll technology on the site Real Doll.

Here Bennato raises an important issue. This is not the first time that a man falls in love with an inanimate object. One can fall in love with a natural object (a landscape, a land, a river, a lake, a mountain, etc.) or with an artificial object (a house, a car, a painting by Leonardo, a musical instrument, a composition, etc.). The problem is that, unlike what happens in the case of the love for a living subject (another human being, an animal), the love for an inanimate object is unreciprocated. Therefore, it remains short of something and incomplete compared to a traditional relationship. But what will happen if and when the robot becomes or seemingly becomes alive? In this case, a human being might prefer a relationship with an artificial being. But this relationship might be even stronger and more exclusive than a traditional relationship, because the robot, like Galatea, would represent the idea of perfection dwelling inside the creator's mind. It rarely happens that we meet someone whom we deem to be 'perfect,' 'ideal.' But what will happen when we can make him, or her, or it?

This problem has been explored in science fiction. So, we could say that sexual and love relations between humans and artificial beings represent a truly science fictive theme. Think about the love story between NDR 114 and Portia in *The Bicentennial Man*. Or about the relationship between android hunter Rick Deckart (Harrison Ford) and the replicant Rachel in *Blade Runner*. In the film the relationship is romantic, but in the original novel by Philip Dick, that inspired the movie, a crude physical and sexual intercourse takes place, with the female android as dominant. But I would like to remind that Italian writer Ippolito Nievo has fun imagining a science fiction world in which men would make love to robot-women. And robot-women would be banned because they would entail an existential risk for society (Campa 2004). His *Storia filosofica dei secoli futuri* [A Philosophical History of Future Centuries] is published in 1860. So, this idea runs across the whole of history, from Greek mythology to the present.

The novelty is therefore not the idea to generate the perfect partner, but the fact that today the realization of this Faustian dream appears possible. The prototypes of artificial sexual partners produced today perform worse than humans. Nevertheless, for people who have no partner, they may shortly become an acceptable alternative to masturbation, prostitution or to other forms of sexuality. Today many people engage in sex with objects. As machines get better it will become ever harder to call this kind of relationship 'fetishism.' We can then imagine a moment when the robots, now nearly indistinguishable from humans, but much more handsome, will seem conscious. Or even be conscious.

Some scientists have paid serious attention to the hypothesis that might emerge a model like Galatea, a robot able to make one fall in love with it. The mystical physicist Frank Tipler, for example, has prophesized that not only could every man have the most beautiful woman that he has ever set eyes on or that has ever lived. Every man could have the most beautiful woman whose existence is *logically possible*. Of course, women will also be able to have what Tipler calls "the logically perfect megamate" (Horgan 1997, 257).[33] We add that these electronic demi-gods could also have flexible roles and appearances, to suit every kind of desire or erotic fantasy.

Tipler regards this scenario as utopic, positive, desirable, paradisiac, and certainly not as a problem. But the love relation between man and machine could of course turn into a moral problem for some people. As our premise states, ethics is in reality relative. On a rigorously sociological plane, the moral problem arises when one group of people regards a certain behaviour as good and another group regards it as bad. Therefore it makes sense to examine what kind of people accept or stigmatize it. Just as the model NDR 114 could become more of a problem to religious people than to atheists or agnostics, so too the Galatea model may become more of a problem to *humanists* (in the sense of speciesists) than for *posthumanists*. If the robot is conscious it is no longer a thing. Why then would it not be allowed the right to love or to be loved?

Evidently this hypothetical sentimental relation could become a moral problem also to the Catholic anthropological vision, because love and sex would thus be uncoupled once and for all from their reproductive function. That the Church might oppose this kind of relationships is not astonishing given that it already opposes homosexual and extramarital relationships. In proposing a negative roboethics, the Church would be perfectly coherent with its own doctrinal position. However, if we analyse the problem from a sociological point of view, then we may ask why Christian anthropology has been successful and why also a non-Christian could subscribe to speciesist theses, discriminating against robots or mixt man-machine couples.

Monogamy, marital fidelity, reproduction restricted to the sphere of the family, all this gives a feeling of security to people who, for physical or emotional reasons find it hard to accept a promiscuous society, one based on free love. In such a libertarian society (this is the way Western society seems to be going), people who are more attractive or intelligent or psychologically robust constitute a perpetual 'threat' to the less endowed, who risk finding themselves condemned to loneliness. A conscious Galatea model could be perceived as a threat by women with mediocre looks or personality. The same would obviously apply to men as

---

33 Tipler, however, is not thinking about machines that are free to move around in the real world, but at a virtual reality inside a 'divine' supercomputer, into which we would be able to upload our minds and live in the form of avatars or simulations. The supermachines, which for a fundamentalist Baptist like Tipler would truly be the incarnation of God, could also resurrect (that is, make simulations of) dead people who have not lived in time to upload their minds into the computer.

soon as the 'perfect' lover could be generated. This category of people might find Christian anthropology an attractive response to their anxiety.

But the opposite might also happen. Finally, the anthropomorphic robot would be a unique opportunity for those people who are least successful in their sexual and emotional relationships: they no longer have to stay single or *be content with the only available partner*; they might at last be able to choose. The problem of loneliness would therefore remain and grow more intense only in the case of speciesists, technophobes, or the inability to afford a robot. The last problem depends of course on the government's or society's political orientation. In Japan the Ministry of Health *gives away* Tama the robot-cat to elderly people as a form of therapy against solitude. Other governments do not even help their citizens with the basics.

In this respect, a remark by Monopoli (2005) is interesting: "One question we should finally ask ourselves is why do we perceive this strong urge in both individual and collective imagination to build robots similar to human beings or to animals. Maybe one answer is that in our relation to others we always fear that they will not accept us, that they will reject us, while we conceive of the robot as obedient and consequently we are certain that it cannot refuse us anything no matter what we do. Somehow, thus, the need to relate to an anthropomorphic robot could be an 'inappropriate' response to our need for relationships, acceptance and recognition by others, and are maybe a herald of new subjective relational experiences and of new forms of objective loneliness."

Monopoli has correctly put the word 'inappropriate' inside inverted commas: indeed it depends on ethical points of view and on the degree of sophistication of the machine. In our opinion, ethical problems arise above all if the machine is not conscious but appears to be. In this case, at an ontological level, we would truly be alone, with the illusion of not being. Love would be a sham. It should be said however that often this also happens in relationships between two human beings. It happens that, after two years of 'love,' eyes are opened and one finds oneself face to face with a stranger, one discovers that one has loved an illusion, a being generated by our imagination and superposed on the real one. Bennato – taking a sociological perspective – has stressed that love is 'real' anyhow if the people involved consider it to be. This is also what the Thomas theorem states: "If men define situations as real, they are real in their consequences." But, on a strictly moral level, the problem of personal isolation, stemming from the illusion that one has a relationship with the perfect partner when in reality (s)he does not exist as a person, cannot be denied.

One last question arises. The robot may appear conscious, or even be conscious without being completely free to choose, or finally it could be totally conscious and endowed with a high degree of freedom. We do not yet know how far technology can go. Human beings as well have varying degrees of consciousness and of initiative or freedom to choose. If it is true that humans dream up and build robots because they need 'slaves,' then they should build machines of the second kind (if this is technically possible): that is, conscious robots, but without initiative or no strong personality. Otherwise there would be no point in making them. An electromechanical demigoddess like the one dreamt up by Tipler, if completely free to choose could tell her master: "I am perfect, but you're not," and leave him. If nothing else, the news of a very expensive robot that walks out on its master would be lethal to the robotic industry, which requires market feedback to be successful. But one could not even buy that kind of being or acquire it some other way because it would be claiming civil rights. Instead, a being endowed with consciousness, but totally dependent (like a pet animal) could be both bought and sold, to the benefit of the maker and the buyer. If we consider this scenario unethical, we should re-examine for consistency the moral lawfulness of how we deal with animals (that incidentally are also destined for the slaughterhouse and not only for entertainment). If instead we regard this as ethical, we could at least raise the issue of the problem of the robo-slave's happiness. It will have to be created such that it is always happy to serve.

4.5.3 The Messalina Model

We could widen the perspective to include the issue of aesthetics and the relation between makers and consumers. To name this model we have found inspiration in Messalina, the Roman noblewoman as beautiful as she was cruel, but a masculine equivalent could be called 'the Saint-Just model.' If the robot is neither conscious nor capable of free will, but simply responds as it has been programmed to, or if the robot is conscious but has well defined psychological traits that limit its possibilities of action and of thinking (as, in any event, happens also to us humans), then an ethical issue might arise: should we *oblige* robotic engineers to establish a clear relation between the physical aspect and the psychological traits of robots? According to Monopoli (2005) the problem can actually be split into two different issues: 1) The emotional resonance of how the robot looks (descriptive roboethics); 2) The ethical and aesthetic coherence (prescriptive roboethics). These are his words:

> Aesthetics, and particularly anthropomorphism or resemblance to animals, is relevant according to the emotional and cultural resonance that it might have on man. We have always been naturally prone to infer additional traits from the appearance; think only of the instinctive association between beauty and goodness that is repeatedly confirmed

in literature at all times, in which we find the figure of the 'kind and beautiful fairy' and that of the 'evil and ugly witch'. The aesthetic element thus takes on a crucial relevance as soon as our mind infers, nearly automatically, our attitude towards who or what we stand in front... A first ethical indication might thus be the coherence between the aesthetic aspect and the 'programming' or 'basic leaning' of the robot. For example, it is ethically correct that an automaton developed to assist people should be good-looking, good, benevolent, while a robot designed for a task that we generally associate with an unfriendly attitude should be aesthetically dull and sullen. This is because the lack of aesthetic coherence sends a false signal, or in other words, a lie that deceives the person and could also inflict emotional damage of a neurotic kind because of the two-faced message that can be derived. In addition, the practical possibility to manipulate as one pleases the aesthetic aspect of the anthropomorphic robot gives rise to an unnatural uncoupling between the interface and the content which could confuse the normal acquisition of the capacity to infer psychological traits from the aesthetic appearance, despite all the exceptions and inconsistencies that we may find in individual people. The general concept is thus that the appearance should not mislead one by inducing incoherent inferences regarding the robot's aesthetic traits.

If the ethical prescription is respected, then the world of robots might indeed function even better than the human world. Effectively we know that inside human social reality we do frequently encounter also the 'ugly good fairy' and the 'beautiful evil witch'. Consistent with this principle, in order to avoid giving rise to misinterpretations, humanoid soldiers ought to be designed and built with an extremely terrifying appearance. It is clear, however, that war ethics is not altogether superimposable on the ethics of every day life. Deception and disinformation are favoured tactics in all belligerent conflicts. In the same way, business ethics is not entirely similar to every day ethics, given that – by analogy – in business bluffing is often as admitted and tolerated as in a game of poker (Carr 1968). While on the contrary a friend, a relative, a fiancé would never forgive bluffing or deception intent on harming them to our advantage. This is why it should not surprise us to see many Messalinas on the battlefield or involved in intelligence activities.

4.5.4 The Gurdulù Model

Another issue we need to assess and discuss is the possible mutual imitation between humans and automata. I will name 'Gurdulù Model' the robot uncritically aping human behaviours, in virtue of the psychological traits of a character in the novel by Italo Calvino *The Nonexistent Knight* (1962). Gurdulù is the squire of Agilulfo, the non-existent knight, and his exact opposite. Agilulfo does not exist physically (only his armour does), but he is endowed with a very strong resolution and personality: nothing deflects him from his goals. His squire on the contrary has a body, a physical consistency, but he has no will: he empathizes with all he sees. A robot that obediently follows a program will be like Agilulfo, resolute when also mistaken. To overcome this problem we are building robots able to learn via imitation, exactly like children. But the robot programmed to imitate humans, instead of acquiring a personality, might remain a kind of Gurdulù, and give rise to the exact opposite problem. In this case we would be confronted with a technical problem, but not necessarily an ethical one. Unless the robot lives in the vicinity of violent humans.

Of course this assessment depends on social customs, prevailing public opinion and legislation inside the various nations. Nevertheless, in general and abstract terms, we can state that the problem that arises is that of not giving a bad example to robots prone to imitation if we do not want to fall victim to a kind of karmic law. Or, if we want to reconvert robo-soldiers with these traits to house or industrial work, when they come back from war, the problem of their 'ethical reprogramming' arises.

But the Gurdulù effect could apply also in the opposite case. We humans might begin to imitate the behaviour of our robots. The problem arises especially in the case of children, who by nature are little Gurdulùs, without a completely formed personality. Monopoli (2005) remarks: "One interesting aspect that is likely to emerge is the influence of anthropomorphic robots on human development. Children have always been taken care of by others who, whatever their intention, always play the part of teachers and role models. The option of entrusting robots for childcare should therefore be evaluated by taking into consideration also the interests of the child and of humankind." We can take it that the problem will not have an easy solution.

4.5.5 The Golem Model

Finally we arrive at the main problem of futurological roboethics – an evolving problem that has inspired so many sci fi movies, from *Terminator* to *The Matrix*: machine rebellion. Bennato suggests to define this problem as 'the Golem syndrome.' We have followed his recommendation and added a hypothetical robot called 'the Golem Model' to our list. The emergence of this model should be a major cause for concern to whoever is designing, building and using armies, flocks, and fleets of war robots.

> The bête-noire in the man-robot relation is no doubt the fear that the robot built to perform various tasks (work, protection) will rebel or at least evade control. This fear has been a part of the man-robot relation ever since the earliest days. The very word 'robot' was coined to designate mechanical workers that end up rebelling against their

master. To emphasize how ancient this idea is I like to call it the Golem syndrome, from the name of the creature in Hebrew Yiddish mythology, made famous by a short story by Gustav Meyrink, in which one tells of a clay creature (the robot) created by the rabbi (the scientist) who gives it life by engraving the word 'truth' on its forehead (the software). The story tells how the Golem, created to defend the Prague ghetto against enemies of the Jewish community, ends up becoming so great a danger to this community that the rabbi has to destroy it. Do we need to worry about autonomous machine running out of control? Maybe yes. DARPA's budget includes a project to fund 5 different projects that may give life to a robot for military use: war zone recognisance robots, smart dust, that is, nanorobots that when scattered over the territory change into intelligent sensors and so on. It is too early to speak of the terminator, but I think that this is a matter of time, not of technology (Bennato 2004).

So Bennato admits that the worries expressed by sci fi writers are not so far-fetched. And we think that he is right. There are real risks to humans and to other forms of biological life that derive from the possible unforeseen behaviour (or from programming by unscrupulous humans) of humanoid robots or inorganic machines in general. It may therefore be rational to talk today about what we might expect from our artificial children, to avoid that the possible collateral undesired effects result in a total ban on humanoid machines, meaning forgoing everything, other than the risks, that these machines will have to offer.

5. Soldiers and Automata

5.1 Defining robotic weapon

Given that military robotics is a relatively recent phenomenon, a conventional agreement upon terminology does not yet exist. Therefore, the preliminary praxis in every scientific work, namely to clarify the terms and concepts, is even more necessary in the present context. In US military and political circles – one of the most active nations at the cutting edge in the development and use of these technological products – the term-concept 'unmanned system' has been introduced to denote systems of weaponry that do not require the presence of human beings where they are located. Such systems are piloted (remote-piloted) *at a distance* by human beings, and even – in the most evolved systems – endowed with greater or lesser autonomy to decide and act. So they are referred to as 'unmanned systems' to distinguish them from 'manned systems,' that is systems without a human operator as distinguished from systems with a human operator. In addition, journalists prefer to use more suggestive expressions such as 'war robot' or 'robot soldier,' even if on closer examination these terms are only used to refer to the more advanced and therefore controversial 'unmanned systems,' that is, those that are of some interest to the press.

In this work we have decided to use the expression 'unmanned system' (UM) as the generic term to refer to any systems of robotic weapon with a military use. We also regard the expressions 'military robots' or 'robot weapons' as being literally equivalent to UM, while the term 'robot soldier' refers only to especially advanced weapons systems, the kind that have some decision-making capabilities, and built for authentic combat.

For a long time, the United States have been compiling and making public a collection of documents with the title *Unmanned Systems Roadmap* that takes stock of the situation on the features and uses of the military weapons currently available to the army and tracks the future development of these weapon systems over the next twenty-five years. So we have the *Roadmap 2005–2030, 2007–2032, 2009–2034, 2011–2036, 2013–2038*. The last three versions have been called 'Integrated' because they attempt to integrate the different aspects of the construction and the use of military robots from the point of view of their interoperability. Priority was given to independent accounts and blueprints of the different typologies of military robots that were worked out and then 'added together.'

The *Office of the Secretary of Defense Unmanned Systems Roadmap (2007–2032)* does not give a precise definition of *unmanned systems*, but a definition of an *unmanned vehicle* – the element that constitutes its main component – hints at the meaning. Here is the definition proposed by the document:

> Unmanned Vehicle. A powered vehicle that does not carry a human operator, can be operated autonomously or remotely, can be expendable or recoverable, and can carry a lethal or nonlethal payload. Ballistic or semi-ballistic vehicles, cruise missiles, artillery projectiles, torpedoes, mines, satellites, and unattended sensors (with no form of propulsion) are not considered unmanned vehicles. Unmanned vehicles are the primary component of unmanned systems (Clapper, Young, Cartwright, and Grimes 2007, 1).

So, as well as a positive definition, the vehicle is also given a negative definition, which rules out a whole range of technological products used in war: ballistic vehicles, missiles, artillery projectiles, torpedoes, mines, satellites, static sensors. Positively, these vehicles are ones with their own type of propulsion, that leaves out the human operator, that can act autonomously or be remote controlled, can be reused many times, and can carry a lethal or nonlethal load. They can in fact carry surveillance systems (video cameras, radars, sonars, microphones, etc.) or lethal weapons (cannons, machine guns, missiles, rockets, etc.). The system of a military weapon is defined by the entire vehicle – its form, propulsion system, dimensions, weight, velocity, etc. – and by the load it carries – its electronic brain, its sensors, its weapons, etc. – that together define its belligerent function.

5.2 Robots of the sky, the sea, and the land

The various editions of the *Unmanned System Integrated Roadmap* offer a large catalogue (albeit incomplete) of robotic weapons systems. Mind that we will not speak about the technical features of every single model, but only of the best known ones. Best known since they have had the honour of media attention precisely because they are ethically controversial in some way or other.

To begin, unmanned systems are divided into three major groups depending on where they are being deployed: in the air, on land, in water. We therefore have unmanned systems equipped for air warfare (UAS – Unmanned Aircraft System), for ground warfare (UGV – Unmanned Ground Vehicle) and for naval warfare (UMV – Unmanned Maritime Vehicle). The latter subdivide in their turn into two categories: Above Water (USV – Unmanned Surface Vehicle) and submarines (UUV – Unmanned Undersea Vehicle). Researchers have renamed UAS as 'flying robots' or 'drones', a term whose origin is related to the shape of these aircrafts (Whittle 2014, Sloggett 2014, Rogers and Hill 2014).

Also the press has noticed the proliferation of these military robots, as a recent report in an Italian daily attests: "Bang, a target is hit, no soldiers employed. This is the evolution of defence systems that on-going wars do much to accelerate. Recognition, attack, transportation, tracking and rescuing are tasks more and more frequently given to robots, which paves the way for the automatized warfare prefigured in science fiction movies. Under the generic appellation of Unmanned Systems, these weapons, that function without a human pilot, were first in use in aviation and have now been fitted inside motorboats, helicopters and motor vehicles" (Feletig 2010).

The article reports that the first 'drone' was used by Israel in the Yom Kippur war, and that sixty years of incessant warfare combined with a cutting edge high tech industry have made Israel the leading nation in the development and production of unmanned weaponry, "surpassing the gutsy US military industry and the land of robotics, Japan." For the sake of precision, it is necessary to recall that "remotely piloted aircraft first appeared during World War I" (Zaloga 2008, 4). Remote controlled aircraft were also used by the Americans in the forties, when they tested the effects of the first nuclear bombs. This of course does not intend to diminish Israel's remarkable technological work in the field.

So the article continues: "During the first Gulf War, in 1991, the Air Force had about a hundred drones; today it is deploying 7000 and keeps churning out new models in response to a demand that knows no limits. This race against the clock is to blame for the high number of accidents: 100 times the number of those involving manned aircrafts according to a study by the Congress. 2009 was the year of the watershed: US aviation trained more pilots in front of a screen with a joystick than in the cockpit holding the control stick. Fewer victims, less expensive to train, but surely more frustrating for Top Gun aspirers."

The article states that there are about forty nations that are developing UM technology. As regards the Italian contribution, it mentions the efforts by Alenia Aeronautica, holder of the patents of SkyX and of SkyY, in addition to taking part in the nEUROn program for the construction of a European unmanned military aircraft and in the Molynx program, the goal of which is the development of a high-altitude robotic twin-motor with up to 30 hours autonomy.

The main goal of the revolution of unmanned vehicles, on land or in the air – the article continues – is that of decreased risk to soldiers. But it is also to contribute, a little like satellites in space, to the complex virtual network of sensors and communications that extend across the stage of operations. Add to this considerations of an economic nature: the take-down of a drone, that flies in any weather, is the equivalent of throwing 45 millions dollars down the drain, if it is a jet fighter 143 millions, naturally not counting the human loss. The US armed force aims for the creation of a fleet of unmanned vehicles equal to a third of the total before 2015. Market valuations predict that turnover in the UM sector may reach 5 billion euros in Europe between 2010 and 2020, and double in the ten years after that and arrive at a total global level of 45 billion euros by 2030.

The article clearly shows one important implication of this new robotic arms race: even if Western nations are at the forefront today, military robots are not the prerogative of these nations, and everything leads one to think that in the future wars will be fought more and more exclusively by machines. More than ever before they will be wars of technology and of industrial systems. Indeed, guerrilla warfare also bets on the potentialities of military robots, so much so that in the Lebanese conflict of 2006 Hezbollah launched 4 drones, the fruit of Iranian technology, on Israeli locations.

Finally, one should keep in mind all the negative and positive repercussions (depending on one's point of view) that the development of military technology has always had on civilians. Even when they are conceived of as systems of weaponry, drones are not limited to military uses. Unpiloted aircrafts are used for the relief work in the case of natural catastrophes and to enforce law and order. For example, the US Coast Guard uses them. New York Air National Guard navy is endowed since 2008 with Predator, an unmanned aircraft nine meters long already used in the war in Yugoslavia. Some models were also in use in Italian skies, on the occasion of the G8 Summit held in Aquila. Remote controlled aircrafts surveyed the crowds spotting turmoil or demonstrators who tried to break into the red zones. "Also the KMax, the unmanned helicopter of Lockheed and Kaman, is increasingly used to transport gear for the troops, as well as for the transportation of special goods to high altitudes and to intervene in forest fires" (Feletig 2010).

Yet the article by *la Repubblica* mainly focuses on Israel. According to this newspaper, 10 or 15 years from now at least a third of the vehicles in use by the armed forces will consist of UM.

Guardium, an armoured vehicle designed by GNius to patrol the borders with Lebanon and Gaza, came into use at the beginning of 2009. It is a small Jeep, similar to a golf cart, fitted with completely automatic command, control and navigation systems. Since last year civilians and the army in the occupied territories have begun using remote-controlled bulldozer convoys to resupply. Rexrobot, a six-wheel vehicle with the carrying capacity of 200 kg of goods to follow the infantry susceptible to receive and execute vocal commands is currently undergoing evaluation in the Israel Defence Forces. Soon will be launched high-velocity unmanned vessels designed by Rafael Industries, with a rigid shell and an inflatable cockpit. The motorboat Protector USV, called Death Shark, is equipped with 4 ultra high definition panoramic cameras (which can capture details 10 miles away) able to shoot in 3D, sonar systems, electro-optical sensors, and remote laser-controlled machine guns able to fixate the target even in rough sea.

What these machines contribute to dangerous demining operations is also fundamental. Many robot models have been designed to explore mined areas and to spot the contrivances. Since mines too evolve – for instance they are now made of synthetic materials that escape the metal detector – the robot's sensory apparatus must similarly evolve to spot these lethal contrivances. For example, "there are the mine sniffers, that the robotics laboratory of the University of Brescia is currently working on, that use 'artificial noses' pegged to entirely autonomous structures that will recognize the smell of the explosive just like dogs. Researchers in Lausanne have tackled the problem of rough terrain by equipping the mine-seeking robot with mountain bike style wheels fitted with crampons to 'escalate' rocky ground. Today some models even work on solar power" (Feletig 2010).

The picture given by this newspaper, even though not always precise and even though it deals exclusively with the Middle East, is detailed and informed enough. Reading the *Roadmap* by the American Department of Defense tells us that the United States pursue the goal of the robotization of the armed forces with a determination no lesser than that shown by Israel. Innumerable prototypes and models are (or have been) produced and used. Here we shall limit ourselves to giving a few examples of each type, in order to give a feel for the technological level that has been reached or that one wants to reach in the future.

5.2.1 Sky Robots

"An unmanned aircraft system (UAS) is a 'system whose components include the necessary equipment, network, and personnel to control an unmanned aircraft.' In some cases, the UAS includes a launching element" (Winnefeld and Kendall 2013, 4).

As regards robotized military aircrafts, one model that unquestionably deserves looking into is the MQ-1 Predator, produced by General Atomics Aeronautical Systems, Inc. In use by all three American armed forces, in 2007 120 specimens were delivered, 95 available and 170 commissioned. Since 1995 the Predator has completed missions of reconnaissance and surveillance in Iraq, Bosnia, Kosovo and Afghanistan. In 2001, the US air force fitted Predator with a laser designator to guide ammunition with high precision and enabled it to deploy Hellfire missiles. As a result of these modifications, the machine became multifunctional, that is capable of both combat and reconnaissance. The upgraded version (MQ-1) completed 170,000 flight hours (as of July 2006), of which a good 80% had been in combat. Today the machine has been taken out of service.

Various 'successors' or models developed from the Predator have already been produced by the same company. One of these is the MQ-9 Reaper. In 2009, the inventory of the *Roadmap 2009–2034* states that 18 specimens have been delivered and 90 planned. A few years later, the *Roadmap 2013–2038* confirms that 112 vehicles of this type are in service (as of July 1, 2013), and provides a detailed case study of this machine which "illustrates the strategy and actions required, when proper initial lifecycle sustainment planning was not done, to transform the sustainment of unmanned systems from a short-term, rapid-fielding environment to a long-term sustainment environment" (Ibid., 142–144).

The MQ-9 Reaper is a robotic aircraft able to operate at medium altitude, with very high flight autonomy (up to 24 hours). As regards the mission, the priorities have been reversed. This system is primarily a hunter-killer system for critical targets, thanks to electro-optical devices and laser-steered bombs or missiles, with only a secondary role given to the system used in intelligence, reconnaissance and surveillance.

One of the systems adopted by the USAF for high altitude reconnaissance and long flight autonomy is the RQ-4 Global Hawk by Northrop Grumman Corporation (12 machines delivered and 54 planned in 2009, 35 in service in 2013). It is capable of monitoring an area of 40,000 nautical square miles per day, at a maximum altitude of 65,000 feet and with autonomy of up to 32 hours. Surveillance is entrusted to very advanced systems, first tested in 2007: Advanced Signals Intelligence Program (ASIP) and Multi-Platform Radar Technology Insertion Program (MP-RTIP).

Nevertheless, if the principal requirement is to keep the 'spy' flying for many days, without needing to return to base, even for daily refuelling, as is the case for Predator and Global Hawk, the aerostatic robots offer the best performances. Currently the army uses some RAID (Rapid Aerostat Initial Deployment), with a flight autonomy of five days and able to reach an altitude of 1000 feet. This model was used in Afghanistan with decent results. However, much more sophisticated aerostats are being built, such as the JLENS (Joint Land Attack Elevated Netted Sensor), fitted with radar and sensors, able to keep flying at 15 feet for 30 days. Twelve specimens of this model have been planned in 2009. Or the revolutionary PERSIUS of HUAV (Hybrid Unmanned Aircraft Vehicle) typology, manufactured by Lockheed Martin Aeronautics, fitted with sophisticated sensors, capable of flying for three weeks at 20,000 feet without returning to base, and able to move with a hybrid propulsion system.

Other 'flying robots' have shown themselves to be particularly useful to the armed forces because of their small dimension and their easy launch and recovery. In this category we find: small gunships like the Wasp by the AeroVironment, of which almost one thousand specimens have been manufactured; micro-mini aircrafts like the RQ-20 Puma (1137 specimens in service in 2013) or the RQ 11

Raven (7332 specimens in service in 2013); and remote controlled helicopters like the XM 157 Class IV UAS, with 32 specimens provided for the Brigate Combat Team in 2009.

The most futuristic model of robotic aircraft the *Roadmap* mentions is no doubt the X47B by Northrop Grumman Corporation, still at the prototype stage and belonging to the category of Unmanned Combat Aircraft System. Its shape is reminiscent of the interceptor ships of the TV series Battlestar Galactica, and so much so that one might mistake them for an alien spaceship. Only this time the UFO does not contain green men, or men of any other colour. Its captain is the grey matter of the on-board computer. It must be able to take off both from runways and from aircraft carriers, to fly at an altitude of 40,000 feet with 9 hours autonomy, and to carry weapons and bombs of reduced diameter.

Its first ground flight took place took place at Edwards Air Force Base, California, on 4 February 2011. As we read in the Northrop Grumman's website, "[i]n 2013, these aircraft were used to successfully demonstrate the first *ever* carrier-based launches and recoveries by an autonomous, low-observable relevant unmanned aircraft. The X-47B UCAS is designed to help the Navy explore the future of unmanned carrier aviation. The successful flight test program is setting the stage for the development of a more permanent, carrier-based fleet of unmanned aircraft" (Northrop Grumman 2015).

Italy as well has a tradition of designing and building robotic aircraft. Feletig (2010) only mentions the Sky-X and Sky-Y by Alenia, but the Falco, manufactured by SELEX Galileo and designed by Galileo Avionica, certainly also deserves to be mentioned. This model is presently at an advanced experimental phase. It is a small size tactical aircraft designed for reconnaissance and surveillance. Its first flight took place in 2003, but the machine has been officially in service since 2009. Even though the SELEX has not rendered public the name of the user, one knows that five systems (a total of 25 aircrafts and corresponding ground control systems) have been sold to Pakistan. In August 2009 the UM Falco was launched using a Robonic MC2555LLR catapult and has completed the test flight. The first flight by aircrafts fitted with high resolution radar and sensors called PicoSAR (synthetic aperture radar) took place in September the same year. In August 2013, the Selex ES Falco was chosen by United Nations to be deployed in the Democratic Republic of Congo "to monitor the movements of armed groups and protect the civilian population more efficiently" (Nichols 2013). The Falco flies at 216 km/h and can reach a height of 6500 meters; it is 5.25 meters long and weights 420 kilograms. It is not designed for combat, but a model called 'Falco Evo' fitted with weapons is currently being studied.

5.2.2 Sea Robots

Unmanned Maritime Systems (UMS) "comprise unmanned maritime vehicles (UMVs), which include both unmanned surface vehicles (USVs) and unmanned undersea vehicles (UUVs), all necessary support components, and the fully integrated sensors and payloads necessary to accomplish the required missions" (Winnefeld and Kendall 2013, 8).

As regards military robots operating at sea, above or below water, the main mission would seem to be mine hunting. There exists a whole collection of submarines with a shape and propulsion engine similar to those of a torpedo, but fitted with a 'brain' and sensors. The primary task of these machines is the ability to spot mines from among other objects, also taking into account the difficulties specific to marine environments that differ from conditions on land.

Relevant companies are fiercely competing to produce the prototype whose performance will ensure their leadership. Statistics are used to detect the object correctly. Still today it happens that all sorts of objects are mistaken for mines or, worse, that genuine mines are not recognized. We shall not give a lengthy description of the technical features of these machines, but confine ourselves to mention one submarine and one surface vehicle.

Amongst the Unmanned Undersea Vehicles one may take note of the Swordfish (MK 18 Mod 1) by Hydroid LLC, a company that is particularly active in this sector. As for surface vehicles, one example is the MCM (Mine Counter Measures) by Oregon Iron Works, currently in the experimental phase. Surface vehicles for other tasks are also being designed, such as the ASW USV, whose function is revealed by its name: Antisubmarine Warfare Unmanned Surface Vehicle; or the Seafox, an unmanned motorboat specialized in coastal surveillance and patrolling.

5.2.3 Land Robots

Land robots or, more precisely, Unmanned Ground Systems (UGS) "are a powered physical system with (optionally) no human operator aboard the principal platform, which can act remotely to accomplish assigned tasks. UGS may be mobile or stationary, can be smart learning and self-adaptive, and include all associated supporting components such as operator control units (OCU)" (Winnefeld and Kendall 2013, 6).

The main mission of land military robots is to clear the ground of mines and explosive devices that are a true nightmare for the Allies' soldiers in Iraq and Afghanistan. Because of their widespread use in this field the MTRS (Man Transportable Robotic System) MK1 and MK2, produced by i-Robot Corp. and by Foster-Miller Inc. respectively, should be mentioned. The *Roadmap 2009-2034* reports that a good 1439 specimens of these machines are already found on the battlefield, but the goal is to roll out 2338 in the coming years. These very useful machines detect and neutralize the explosive devices that military contingents encounter on their path. On the battlefield 324 MK3s by Northrop Grumman Remotec and 1842 MK4s by Innovative Response Technologies are also in use. These are budget robots that save the lives of a great number of people.

While deminers have been widely used for a long time, the same cannot be said of combat robots (the so called TUGV – Tactical Unmanned Ground Vehicle), that is, of machines with no human operator, that are capable of attacking and killing human beings. Various prototypes are currently studied. One of these is Gladiator, of which six specimens have been produced by Carnegie Mellon University for the Marine Corps. Gladiator is an armed and armoured combat robot, endowed with a collection of sensors and weapons that include: infrared sensors, video camera, rocket launcher and machine guns of type M240 and M249. The vehicle moves on wheels, can be remote controlled by a soldier up to one nautical mile away and is equipped with a system that hides the exhaust gas.

Another machine destined for combat is the Armed Robotic Vehicle (ARV) by BAE Systems, and produced by the US Army. 679 of these have been commissioned in 2009. It weighs 9,3 tons and has been designed to perform two specific tasks. The first is reconnaissance: indeed, the ARV-RSTV model (Reconnaissance Surveillance Targeting Vehicle) is able to scan an area and find, detect and reconnoitre targets with great precision, thanks to its sophisticated on-board sensors. Instead, the ARV-A model is fitted with a range of lethal weapons, among which a medium-calibre cannon, a missile launching system and machine guns. Once the experimental stage is completed, it will be possible to use this model in combat.

However, ground warfare has come to a halt. Among the many reasons one can list the misfortune that happened to Forster-Miller's SWORDS. This is a tiny caterpillar robot carrying a light M249 machine gun. The press and the manufacturer give different accounts, but it would seem that the robotic weapon did not behave as it was supposed to.

On April 11th 2008 *The Register* published a gloomy headline: "US war robots in Iraq 'turned guns' on fleshy comrades." The author tells how the robotic vehicle began to behave unpredictably, stopped obeying orders and spread panic among the soldiers. The tone varies from ironic to apocalyptic: "American troops managed to quell the traitorous would-be droid assassins before the inevitable orgy of mechanized slaughter began… the rogue robots may have been suppressed with help from more trustworthy airborne kill machines, or perhaps prototype electropulse zap bombs" (Page 2008).

The news was followed above all by *Popular Mechanics,* which interviewed Kevin Fahey, the US Army program executive officer for ground forces, about this incident. He confirmed it and explained that the robot began to move when it was not supposed to move and did not fire when it was supposed to fire. No human was wounded, but the drill was stopped from precaution. The officer added that "once you've done something that's really bad, it can take 10 or 20 years to try it again" (Weinberger 2008).

In reality, in a later article, also published by *Popular Mechanics,* Fahey explained that the SWORDS war robots are still in Iraq, and that they have been neither destroyed nor withdrawn. Cynthia Black, Foster-Miller's spokesperson, also wished to explain that "the whole thing is an urban legend" (Weinberger 2008). Black clarified that it is not a self-driving vehicle. That it can therefore not fire unless told to do so. That the uncommanded movements were due, not so much to the computer going awry, but to a trivial mechanical problem. The robot was put on a 45-degree hill and left to run for two and a half hours, and the motor overheated. When this happens, the engine automatically switches off to avoid breakage. But because it was on a slope the machine started to skid, and gave the impression of autonomous movement. This is the producers' version. Fact is that three SWORDS war robots have really stayed on the battlefield, but placed in fixed positions. Some senior official even wondered if it would not be more practical to put the machine guns on tripods.

So one is given to understand that a hypothetical slowing down of the experimentations is not due to this trivial incident, but to a much more important structural situation, such as the economic crisis that has plagued the United States for the past years and the contextual withdrawal of US troops from Iraq, expected by the end of August 2010 and completed 10 days ahead of schedule.

Experiments continue in Afghanistan and in laboratories. Also under study is an Unmanned Ground Vehicle, able to spot chemical, biological, radiological and nuclear (CBRN) devices. The iRobot is in fact designing this kind of machine for the US Army.

But the robots on the battlefield can also reveal themselves useful, not only for observing the enemy, unearthing and detonating bombs or fighting. They can also massively assist the wounded during belligerent missions. Many times the wounded cannot be recovered or cured, and therefore they die from blood loss or from the wounds they have incurred, because the place where they find themselves is out of reach or under enemy fire. Here is a machine that can carry out this delicate and dangerous task instead of the stretcher-bearers or of machines operated by humans. Applied Perception Inc. has produced a prototype of Robotic Combat Casualty Extraction and Evacuation. In reality, it is a robot couple. A 'marsupial' vehicle serving as ambulance and connected to a vaguely humanoid machine, with mechanical arms, serving as paramedic. The vehicle is endowed with laser, radar, sensor and systems of navigation that permit it to avoid obstacles and to reach the location of the wounded. In addition the machine is endowed with a telemedicine audio-video system that allows the patient to communicate remotely with a doctor.

The press tells us of other innovative projects that might become a reality in the future for military and civil personnel. All the UGVs mentioned in the *Roadmap* are endowed with wheels, because it does not yet seem that humanoid bipedal models are combat-ready. However, it seems only a matter of time. The performances by the Boston Dynamics 'quadruped' called Big Dog are indeed astounding. In a Fox News footage, Matt Sanchez (2009) describes it as follows:

> Using a gasoline engine that emits an eerie lawnmower buzz, BigDog has animal-inspired articulated legs that absorb shock and recycle kinetic energy from one step to the next. Its robot brain, a sophisticated computer, controls locomotion sensors that adapt rapidly to the environment. The entire control system regulates, steers and navigates ground contact. A laser gyroscope keeps BigDog on his metal paws — even when the robot slips, stumbles or is kicked over. Boston Dynamics says BigDog can run as fast as 4 miles per hour, walk slowly, lie down and climb slopes up to 35 degrees. BigDog's heightened sense can also survey the surrounding terrain and become alert to potential danger. All told, the BigDog bears an uncanny resemblance to a living organic animal and not what it really is: A metal exoskeleton moved by a hydraulic actuation system designed to carry over 300 pounds of equipment over ice, sand and rocky mountainsides.

This robotic animal cannot fail to attract the attention of the Italian press. Fabrizio Cappella (2008) writes in *Neapolis* that "it seems midway between a dog and a giant spider: it has four legs, no head and it walks on broken ground across obstacles: it is called Big Dog and its video has set ablaze the imagination of internet surfers who, for some time now, have fired the wildest comments at the bizarre creature." The article reveals that this is a project funded by the Pentagon, and that its full name is "Most Advanced Quadruped on the Earth."

Effectively, Big Dog's locomotion is surprisingly natural also on very rough terrain, where it manages to keep its balance in the toughest situations, for example after it has been kicked or after having slipped on ice. The robot moves at about 4 mph and its frame is made of steel: hidden inside, in addition to the petrol engine, are a computer, sensors, video cameras and a global positioning system. It is capable to transport hundreds of kilos of gear and can withstand collision with wheeled vehicles and caterpillars. Its purpose is military; one studies its usefulness to troops in warzones, its ability to carry heavy loads and to transport the wounded. The Pentagon appears to have great faith in the success of the project, given that it has invested 10 million dollars in the prototype. Now in its second version, Big Dog will be further developed and its definitive version ought even to be able to gallop thanks to the form of its legs that are very similar to those of racing animals.

The comments found online are divided. Some are enthusiastic and others admit to being intimidated and concerned. Here two fundamentally opposed ethical leanings play an important part: on the one hand technophilia (the pride of belonging to the human species, able to build these wonders), on the other technophobia (the refusal to give up pre-industrial or pre-neolithic lifestyles). What is certain is that this machine, whose locomotion is so similar to that of living beings, does not leave one indifferent.

Another machine that thrills the imagination and stirs up discussion among journalists and readers is a robot called EATR (Energetically Autonomous Tactical Robot), not because of the weapons it carries or because of its locomotive capacities, but because of its system of propulsion and fuel supply. Here it would be appropriate to use the term 'feeding,' which refers as much to machines as to living beings. Effectively the EATR is fuelled much like a human being.

Riccardo Meggiato (2010) writes in *Wired*:

> Don't be scared if one day, pretty soon, you see a robot among the grazing cows: robots also eat, didn't you know? The EATR, an acronym that does not refer to some exotic train but stands for *Energetically Autonomous Tactical Robot,* is a model that feeds itself: it literally eats plants and converts them into biofuels that it uses to move. The ultimate purpose of this project, still under development, is to create vehicles that cannot only do without classic fuels, but are also able to provide for their own energetic needs. The EATR is the work of *Robotic Technology* in Washington, and its development is funded by *Defence Advanced Research Projects Agency* (aka DARPA). All right, it's the army longing to create autonomous military vehicles, but it is also clear that this kind of technology, once in place, could benefit many different sectors.

Wired also reveals some of the technical features of the feeding-propulsion system.

> So how does this cool gizmo work? Oh, it's easy: it forages plants with a mechanical limb and ingests them into a combustion chamber. Once on, it generates heat that warms up reels filled with deionized water, which evaporates. The steam obtained then activates the six pistons of a special engine, which activates an energy generator. This one, finally, is stored in specific batteries and used if needed. To give the system more autonomy, researchers at Robot Technology have developed a range of recovery solutions. For example, if steam escapes from the pistons it is promptly condensed, turned into water and sent back to the combustion chamber. And if there is a shortage of grass to graze, no worries: EATR can happily run on traditional fuels as well, like diesel, petrol, kerosene, and even cooking oil. This, cows can't do.

So the robotic system promises a solution to one of the major problems that Unmanned Ground Vehicles encounter: poor autonomy. In order to function on the battlefield, sometimes far from provision lines, it is important not to be constrained by matters of energy. An electric battery may be enough for a vacuum cleaner or a lawnmower, but its performance is unlikely to do for a robotic soldier lost in the Afghan mountains.

Robert Finkelstein, the boss of Robot Technology, guarantees that 68 kilos of plants make the EATR energy autonomous for about 160 km. The vegetarian engine has been construed by *Cyclone Power Technology* from a design by the research centre in Washington. The first experiments predict their integration into a Humvee type military vehicle. Whether it will be mass-produced will depend on the test results, but the producers obviously hope to send the EATR to the battle scene as soon as possible.

Hence Meggiato concludes: "EATR applications are manifold and go beyond the military. While some have ironically pointed out that they can also be used as weapons to gobble up enemy troops, others view them as tractors ready to work non-stop without refuelling, possibly controlled by another robotic system that does not need human intervention. In the end, whether it is a Terminator or a Winnie the Pooh in high tech wrapping, the future will be vegetarian."

5.3 The main functions of the military robots

Robots are given the missions that military jargon defines as dull, dirty, or dangerous. In other words, even if some technologies are still not up to replacing man in every relevant task, it does appear rather obvious that the human element is from now on the limiting factor in carrying out certain war missions.

Hard work. Sometimes the battlefield requires work done that the human organism finds it difficult to endure. For example, a human pilot needs a few hours sleep after a long operation; a drone does not. While the longest manned air missions of operation Enduring Freedom lasted around 40 hours, there are now drones that guard some warzones non-stop, remote-controlled by crews on the ground that change every 4 hours. The only limit is aircraft autonomy, but if refuelling can be done in the air then that limit too is removed.

Dirty work. As the *Roadmap 2007–2032* reminds us "[US] Air Force and Navy used unmanned B-17s and F6Fs, respectively, from 1946 to 1948 to fly into nuclear clouds within minutes after bomb detonation to collect radioactive samples, clearly a dirty mission. Unmanned surface drone boats, early USVs, were also sent into the blast zone during Operation Crossroads to obtain early samples of radioactive water after each of the nuclear blasts. In 1948, the Air Force decided the risk to aircrews was 'manageable' and replaced unmanned aircraft with manned f-84s whose pilots wore 60-pounds lead suits. Some of these pilots subsequently died due to being trapped by their lead suits after crashing or to long-term radiation effects." These incidents persuaded the US military to revert to using robots for dirty work.

Dangerous work. Explosive Ordinance Disposal (EOD) is the primary example of dangerous work entrusted to robots. Improvized contrivances found in the streets and in places where soldiers go constitute some of the major threats in the current military campaigns in Iraq and Afghanistan. Coalition forces in Iraq neutralized over 11,100 Improvised Explosive Devices (IED) between 2003 and 2007. A great percentage of these missions was done by ground robots, and the number of UGVs employed in these tasks has skyrocketed: they were 162 in 2004, 1600 in 2005, over 4000 in 2006, 5800 in 2008.

In order that the performances of military robots meet aspirations, commanders on the field at the head of the different armed forces have been asked to submit a priorities list engineers should focus on. Even though the demands of the ground, air and naval armies differ for obvious reasons, it has become clear that they have four common priorities: 1) Surveillance and reconnaissance; 2) Target identification and designation; 3) Counter mine warfare; 4) Chemical, biological, radiological, nuclear explosive (CBRNE) reconnaissance.

Surveillance and reconnaissance. The main priority has revealed itself to be reconnaissance capacity (electronic and visual). For many army professionals information, qualitative and quantitative, is the key element for operational success and robots are the best candidates to gather this information. The ideal robot is able to exert persistent surveillance (or for long periods) on hostile areas, while maintaining some degree of 'covertness.' Robots are promises that the limits of other systems such as manned vehicles, satellites, submarines and unattended sensors will be overcome.

Target identification and designation. The ability to identify and to locate targets with precision in real time is one of the most urgent necessities on the battle stage. It is necessary to reduce the 'latency' and to increase the precision for GPS guided weapons, as well as the ability to operate in high-threat environments without putting warfighters at risk. A quality leap in this sector would improve not only safety, but also be more efficient and efficacious than traditional manned systems.

Counter-Mine Warfare. The most useful yet dangerous mission is that of demining a piece of land or sea. Statistically speaking, since World War II, sea mines have caused more losses of US warships than all other weapons systems combined. The same can be said of landmines and bombs (IED – Improvised Explosive Devices) that are responsible for the majority of losses of the coalition forces in Operation Iraqi Freedom. Commanders regard improving the robot's capacity to find, tag and destroy these devices as a priority. Henceforth robots appear irreplaceable for this sort of work. They have already saved innumerable lives and, as their technology improves, this ought to reduce casualties still further.

Chemical, Biological, Radiological, Nuclear, Explosive (CBRNE) Reconnaissance. The dirtiest of dirty work is that of spotting CBRNE. Yet this kind of weapon of mass destruction also represents the greatest peril for a nation at war. An attack with nuclear, chemical or biological weapons on foreign land or on troops deployed at the front, would have disastrous consequences not just on the waging of the war, but also on the entire military apparatus, on the economy and on foreign policy broadly speaking. Therefore robots are essential, as much to prevent this kind of attack as to observe and monitor areas that have already been attacked, because of their superior sensorial capacities and because of their greater resistance to chemical, radioactive and microbial agents.

In the *Roadmap 2007–2032* the future goals that constructors and users of robotic weapons in military circles have set themselves are the following:

1) "Improve the effectiveness of COCOM [combatant commander] and coalition unmanned systems through improved integration and Joint Services collaboration." To this end one expects new designs and experiments on the battlefield with the most promising technologies, accurately testing prototypes prior to their deployment. Reducing the risk in the use of fully developed technologies is also part of the project.

2) "Emphasize commonality to achieve greater interoperability among system controls, communications, data products, and data links on unmanned systems." Also here the stress is both on security and on safety. On the one hand, it is necessary to improve the 'common control' and 'common interface,' so that the control systems can easily operate the various kinds of robots. On the other hand, it is important to prevent interceptions, interferences, hijacking, so that the enemy cannot take control of these machines and turn their lethal potential against the army owning it.

3) "Foster the development of policies, standards, and procedures that enable safe and timely operations and the effective integration of manned and unmanned systems." These goals include: a) developing, adopting and prescribing commercial and government regulations relative to the design, construction and experimentation of unmanned systems; b) the coordination between the civil authorities that manage the air, sea and land areas for civil usage (the transport of goods and passengers) and the military authorities in order to prevent collisions between manned and unmanned machines; c) the development of

ever better systems of sensors and control, to give robots the necessary autonomy to avoid collisions with traditional means of transportation.

4) "Implement standardized and protected positive control measures for unmanned systems and their associated armament." More specifically, one feels the necessity for a standard architecture common to all unmanned systems, armed or not.

5) "Support rapid demonstration and integration of validated combat capabilities in fielded/deployed systems through a more flexible prototyping, test and logistical support process." More specifically, one intends to develop alternatives to gasoline-powered internal combustion engines, with a particular predilection for high-energy-density power sources (primary and renewable), and if possible common with those of manned systems.

6) "Aggressively control cost by utilizing competition, refining and prioritizing requirements, and increasing interdependencies (networking) among DoD [Department of Defense] systems." In other words, stimulate both competition among manufacturers and their cooperation, while keeping cost reduction as the primary goal.

New requirements were added to this list, punctually recorded in the updated and integrated *Roadmap 2009–2034*. In particular, one can see that the army insists less on control procedures and security standards, and more on the speedy production of the machines and on their necessary autonomy. This is an important change, which in our opinion reflects the fact that, in recent years, robots came to be viewed as more reliable. So we add the two key points:

7) To maintain the sectors of research and development to increase the level of automatization of the systems of robotic weapons, so that they reach the appropriate level of autonomy, as determined by the combatant for each specific platform.

8) Speed up the transition of robotic weapons systems from the sectors of research and development set up by scientists to the hands of the combatants at the front.

It is therefore considered opportune to maximally stimulate the production and use of ever more sophisticated military robots, because of the army's ever more enthusiastic reception and implementation of the robots that arrive on the battle stage. Hence moral uncertainties appear to fade away. Besides, operations like demining and the clearance of explosive device in areas either inhabited or in some way traversed by people, as well as the prevention of attack with chemical, biological or nuclear weapons, will hardly raise any ethical objections. What robots do and will go on doing on the field, in time of war and in time of peace, is nothing other than humanitarian work. The same can be said of aid to the wounded. However, it is true that other questions, such as electronic combat and surveillance, could still raise questions of a moral nature. Add to that man's atavistic fear – symbolically codified in myths, legends and tales – of a rebellion by the creatures against their creator, and one understands that robotic technologies promise to become a main area of applied ethics.

5.4 Main objections to the belligerent use of robots

Given that many look upon war as a negation of ethics (sometimes also when it is defensive), and that technological development itself finds firm adversaries on principle, it is not astonishing that the application of robotics to war has stirred up so much discussion (Relke 2006; Sparrow 2007; Turse 2012; Bashir and Crews 2012; Amhed 2013; Chomsky and Vltchek 2013; Sullins 2013; Evangelista and Shue 2014; Wittes and Blum 2015).

The question however does not engage just pacifists, Luddites, and roboethicists, but also military professionals and engineers. The development of this kind of operation does indeed promise to solve many problems, but it is not without its pitfalls. The debate is therefore more necessary than ever. Here we shall outline the ethical objections to the use of military robotics that we have found most cogent, and, in a second part, we shall evaluate them from an ethical point of view.

5.4.1 Noal Sharkey's plea

A plea by the Royal United Services Institute (RUSI) that denounces the dangers of a robotic arms race and the risk that it would imply for all humanity has caused a particular stir in the media. The plea has made headlines because it is written by experts in the new technologies and, moreover, for so prestigious an institution as the RUSI. Those who are informed about military matters know well that the RUSI is not some hangout of Pacifists or Luddites.

This point of view has found one of its more notable spokespersons in Noel Sharkey, professor of Computer Science at the University of Sheffield. According to him, "the trouble is that we can't really put the genie back in the bottle. Once the new weapons are out there, they will be fairly easy to copy. How long is it going to be before the terrorists get on in the act? […] With the current prices of robot construction falling dramatically and the availability of ready-made components for the amateur market, it wouldn't require a lot of skills to make autonomous robot weapons" (University of Sheffield 2008).

The first argument that the anti-robot front puts forward is therefore the possibility that the enemy could use these creatures against us. Strictly speaking, this is a prudential argument rather than an ethical one. Indeed, it is about our own good, rather than the good of other fellow humans. There is a fear that our own drive for hegemony can turn against us. Western nations are apparently investing huge amounts of money in the construction of these war machines (4 billion dollars in 2010 and a total expense of 24 billion dollars in the case of the United States), but once they fall into enemy hands they are easy to copy. At what point will Islamic fundamentalists or other enemies of the West no longer need kamikaze and suicide bombers, but will be able to direct remote controlled drones with lethal charges against preselected targets? Sharkey has been interested in this problem for a long time, and also worked as an advisor to the BBC during the broadcast of the television series *Robot Wars*.

Maruccia (2010) observes that "the professor does not give much detail as to this presumed facility to build, but he does assure us that a drone equipped with an autopilot guided by Sat Nav currently carries the modest price tag of 250 dollars." He probably refers to mini drones, given that a Predator costs around 4 million dollars, but we can certainly bet that the cost of these technologies will fall substantially. In addition, it is true that mafias and terrorist groups sometimes dispose of large sums of money and that, for the sum that a Nation spends on the purchase of a supersonic jet plane, one can buy 30 Predators.

The second ethical problem that Sharkey brings up is the drones' limited capacity to discern, that is, the possibility of error: because of the 'relative blindness' of the machines currently in use it is not possible to guarantee the discrimination between combatants and innocents or a proportional use of force as required by War legislation. "Allowing them to make decisions about who to kill would fall foul of the fundamental ethical precepts of a just war under *jus in bello* as enshrined in the Geneva and Hague conventions and the various protocols set up to protect civilians, wounded soldiers, the sick, the mentally ill and captives. There are no visual or sensing systems up to that challenge" (Sharkey 2008a, 87). In an article appeared a few months later on *Science*, Sharkey (2008b) clarifies that "even with a definition [of a noncombatant], sensing systems are inadequate for the discrimination challenge, particularly in urban insurgency warfare."

Here the misgiving is mainly ethical, because it concerns others' safety. But let us add that the error could also consist in killing allied soldiers. The so-called friendly fire. Because of this, Sharkey (2007) solicits a serious international debate, one which also takes hypotheses of a moratorium into consideration: "With prices falling and technology becoming easier, we may soon see a robot arms race that will be difficult to stop. It is imperative that we create international legislation and a code of ethics for autonomous robots at war before it is too late." In other words, the international community should evaluate the risks of these novel weapons *now*, rather than sit around and wait while they sneak their way into common use.

5.4.2 Robotic wars as war crimes without criminals?

The question of the possibility of errors is raised also by Andrew Brown in a blog related to *The Guardian*. However, he lays the stress above all on the matter of relieving oneself from the burden of responsibility. Reflecting on the concept of hostile artificial intelligence, Brown warns that the robot has a particular status that it is hard to define: it is not yet a sentient being capable of moral discernment, but neither is it a mere object controlled by man: it has a goal and it pursues it, even though it does not know that. By the way, this is true also for the so-called smart bombs, that follow heat or satellite signals: "The missile, a thing that is both dead but none the less animated by a hostile purpose, violates some primitive expectations of the way the world works. That's one reason it seems so frightening" (Brown 2009).

Brown raises the problem of the moral status of the robot, of the people that are constructing it, that give the order to use it, and that use it. He rejects the idea of those he calls "the protagonists of extreme artificial intelligence," for whom the robot is considered on a par with humans once its behaviour becomes indistinguishable from that of a human (that is, that it passes the famous Turing test). He therefore proposes a further ethical problem linked not so much to the cecity as to the possible lunacy of the robotic soldier. He asks "what would happen to a robot which acted against its programmers' intentions: if it started to shoot everyone less than four feet high, or offer sweets to anything armed with an RPG?[34] The answer is obvious. It would be either reprogrammed or destroyed. A human, on the other hand, would be tried, because a human could be blamed – or praised for what he had done."

According to Brown, an entirely unprecedented problem arises in the presence of hostile artificial intelligence: we could have war crimes without the possibility of identifying for certain the war criminals.

5.4.3 Trivialization and multiplication of armed conflicts

Also Peter W. Singer (2009a) has dealt with this question in a lengthy and detailed article that appeared in *The Wilson Quarterly*. Singer begins by describing the death of a soldier, one much appreciated by his fellow soldiers and by his commander for his courage, tenacity and ability. He had saved many lives but, during a demining operation, the device that he was trying to deactivate exploded, killing him. His comrades in arms picked up his remains and carried them away from the

---

34 *Rocket Propelling Grenade* – a Soviet manufactured anti-tank grenade launching system.

scene by helicopter. When writing their report, the commander lavished words of praise and gratitude for the soldier that offered his life, but said that, at least, there was one thing he was relieved about: "When a robot dies, you don't need to write to its mother."

The death of PackBot cost US taxpayers 150,000 dollars. It will be replaced with few tears shed by its clone. Or by a more advanced model.

Singer starts out with this example to argue that robotic war opens up new sociological, psychological, ethical, legal, and political scenarios. A novelty comparable to that offered by World War I, the first major conflict after the industrial revolution. Drawing inspiration from science fiction writers of the time (H. G. Wells, A. A. Milne, Arthur Conan Doyle, Jules Verne, etc.), farsighted politicians like Winston Churchill and engineers tried hard to put previously unseen 'steel monsters' on the battlefield: armed tanks, aeroplanes and submarines. This brought war to a level it had never reached before. The biggest novelty was that the new weapons (machine guns, gas, armoured tanks, etc.) made a carnage of any attempt to move the front just a few meters, while planes and zeppelins managed to bring the war from the front to inhabited cities and unarmed civilians, and submarines came to threaten passenger ships and unarmed freighters. It radically altered the way in which war was fought, and not just as regards the strictly technical, but also the human.

The same is happening now with robotic arms. Singer, even though he underlines the obvious positive aspects of these weapons for whoever has them, that is, that they spare human lives in one's own faction, he brings up another question for ethical discussion, that someone has called 'the videogame effect.' Those who fight with robotic means are very far from the battlefield and do not always feel as if they were killing living and sentient beings. We could refer to this problem with an expression: 'trivialization of war.'

The testimonies that Singer collects give a fairly clear idea of the new psychological dimension the fighter finds himself in. While the Predator's sensors spot the enemy on the mountains of Afghanistan and attack him with lethal weapons, the human pilot is 7500 miles away in a military base in Nevada. The experience is that of a psychological disconnection between being 'at war' and leading a normal working life. A pilot of Predator describes the sensation as follows: "You see Americans killed in front of your eyes and then have to go to a PTA [Parent Teacher Association] meeting." Says another: "You are going to war for 12 hours, shooting weapons at targets, directing kills on enemy combatants, and then you get in the car, drive home, and within 20 minutes you are sitting at the dinner table talking to your kids about their homework" (Ibid.).

Another interesting question that Singer raises concerns control, human presence in decision-making. We have seen that this is the Euron Codex's first point. Or, to say it like Eliot Cohen – an expert on military questions who has worked in the State Department under the administration of George W. Bush –, "people will always want a human in the loop." Although we may want this, it is time to ask if this is technically possible, if it will not lead to rather paradoxical situations.

In fact, as the number and quality of robotic arms improve, humans will get expelled from the 'loop' little by little. This process was visible already at the time when electronic weapons emerged (radar, radio, sonar, etc.) in the first half of the 20th century (Devereux 1993), and is becoming ever more visible today. Let's begin with an example. During the Gulf War, the captain and radar navigator Doug Fries describes bombing operations as follows: "The navigation computer has opened the aircraft hold door and unhooked the bombs into the dark." Of course other human beings programmed the machines initially, but then one allowed the computer to take over on the battlefield, giving the pilots a merely auxiliary role.

The most tragic event in connection with this kind of procedure took place also in the Persian Gulf in 1988: the case of Iran Air Flight 655. In the eighties US naval ships had been endowed with the computerized defence system Aegis that had four different modalities of action. Among these were the 'semi-automatic' modality, which gave humans the possibility to decide if and what to fire at, and the 'casualty' modality designed to run the ship and defend it if all the men on board were dead. On July 3rd 1988, the USS Vincennes, renamed Robo-cruiser for the Aegis system and because of the captain's aggressive reputation, detected the presence of an aircraft and identified it as an Iranian F-14, and therefore signalled it as an 'assumed enemy'. Although Aegis was set in 'semi-automatic' mode, that is, with the machine given minimum decisional autonomy, none of the eighteen marines and officers of the command wanted to take the responsibility of contradicting the computer. Hence they followed its advice and authorized fire. The missile destroyed an innocent passenger plane with 290 passengers on board, among which 66 were children.

Let us therefore make a list of the errors made: a) The Aegis is designed to oppose the action of Soviet bombers in the north Atlantic in war time and acted according to these directives, and yet it found itself beneath a sky full of civilian planes in peace time; b) His great trust in computers lead the commander to drop a security procedure that envisaged asking higher officials on other war ships for permission; c) once again, deep faith in computer wisdom induced the captain and his collaborators to blindly listen to the advice of the machine, despite the improbable nature of an Iranian attack.

Similar errors have occurred with other robotic or automatic weapon systems. During 2003 the invasion of Iraq, a battalion of Patriot missiles took down two allied aircrafts upon having mistakenly classified them as 'Iraqi rockets.'

Here then is what the situation looks like, beyond the problems. In theory, humans are still in the loop, part of the decision-making, but the truth is that decisions have to be made in seconds, between the computer signal and the possibility of one's own death, and therefore no one feels up to using what now boils down to a 'veto power.' One always allows the robotic weapon to fire and hopes that it will strike the enemy and not unarmed civilians or allies. When acting under such psychological stress, it is as if humans had no role to play.

This situation is summed up in what we could name the 'paradox of controlled autonomy.' Many have become aware of the problem, among these the psychologist and expert on artificial intelligence Robert Epstein: "The irony is that the military will want it [a robot] to be able to learn, react, et cetera, in order for it to do its mission well. But they won't want it to be too creative, just like with soldiers. But once you reach a space where it is *really* capable, how do you limit them? To be honest, I don't think we can" (Singer 2009b).

In other words, one first constructs a machine able to do things humans cannot, and then one still expects that humans would have the last word about what the machine ought to do. This is paradoxical.

The result is that, when releasing thousands of robots on the battlefield, one continuously feels the need to introduce exceptions to the general rule that wants humans to have the last say in all decisions. Let us look at it in more detail, reasoning in terms of degrees of autonomy, and not just in a digital one/zero perspective.

First exception. Just as an official has authority over a certain number of human soldiers, one imagines that an operator could supervise a certain number of robotic soldiers. The problem is that the number of robots that a human being can control is directly proportional to the individual robot's degree of autonomy. To understand the problem, let us imagine that we are playing five videogames at the same time. A Pentagon report stresses that "even if the gunship commander is aware of the position of all his units, combat is something so fluid and rapid that it is very hard to control." In other words, if we really want them to fight, and if we cannot assign one commander to every robot, we have to give them the possibility to respond autonomously to enemy fire.

Second exception. No reminder is necessary that the enemy is as sentient and uses electronic arms just as much as we do. As early as the Tsushima battle of 1905, Russians and Japanese used radio waves to spot their mutual presence or to interfere with the communication between battleships (Devereux 1993, 66–74). If the robotic soldier cannot fire unless a remote operator (a human soldier) authorizes it, then it will be enough to obstruct the communication to render the machines harmless and leave them at the mercy of the enemy. In other words, it makes sense then to set up a plan B in the case communications are cut off which envisages the possibility of robot decisional autonomy. In this case they will be able on their own to defend themselves against threats, hit the enemy and return to the base. We can only hope that they make no mistake.

Third exception. Even if every robotic weapon has its own operator, even if the communication is not broken, even if the enemy does not operate at digital speed, there are situations in combat in which humans cannot react fast enough to neutralize a threat. If a projectile is fired at a robot, it takes a human some time to notice (due to the time of propagation of sound waves, to brain reaction time, to momentary inhibition provoked by noise or fear, etc.), while a robot is at once able to spot the conflagration source and frame it as the target of a laser ray. If one can point a laser at someone who fires, then in the same way one can fire a lethal projectile. That is, if one is working in auto-mode, without waiting for a human operator to give the green light, then one could shoot down anyone firing before he had the time to put away his weapon and hide or run away. It is a very strong argument that soldiers on the field are quick to point out. Which human being would risk life and limb, with a very high probability of instant death, in order to kill a machine? Giving the robot enough autonomy to return the fire would totally change war and guerrilla warfare. It would make armed insurgence pointless, because this one is linked to a need for revenge on the occupying forces. Among other things, introducing this exception could seem attractive not just to soldiers, but also to public opinion, which looks rather favourably at the asymmetry between attacking and responding to an attack (even in a 'superhuman' way). The robots do not aggress humans, but eliminates them if they become aggressive and dangerous.

One is also considering the hypothesis of taking exception to the general rule of control, in partial terms, that is, enabling the robot to fire, but only in order to strike machines and not human beings (other robots, armoured tanks, jeeps, etc.). In this case, a robot could block the enemy by targeting wheels or caterpillars. However, in so doing, the robot would not shelter from enemy fire, given that the human operators would or should survive. And it would not shelter fellow soldiers, given that survivors would keep their ability to fire and kill. The dilemma does therefore not go away, and the idea generally speaking of an exception remains a sensible one.

The problem is that, by multiplying exceptions, one risks giving full freedom to the machines. As robotic weapons become more and more reliable, commit fewer and fewer errors, they will get to so high a degree of reliability and, in combination with the technical impossibility for man to replace the machine, we will reach a point of no return. Let us not forget that, indeed, humans too make mistakes. Military history is rich with episodes of friendly fire being more homicidal than enemy fire. Humans are not less dangerous than computers or robots. Even in the presence of errors, it will be enough that statistics weigh the balance in favour of the computer or the robot, to completely remove humans from both battlefield and decision-making.

What might happen in the future, starting with these observations and looking at the technological trends, it is the emergence of yet another ethical problem: the increase of belligerent conflicts. This at least is the opinion of Lawrence J. Korb, an ex marine officer, the author of some twenty books, who has served also as assisting secretary of defence during the Reagan administration. Korb is a great supporter of robotic weapon systems because these save human lives. However, he is persuaded that this is precisely why technological development will make it ever easier psychologically to decide to go to war. There are two factors that push in this direction, and both are the effect of the automatization of the armed forces: a) The growing disconnection between the military apparatus and civil society; b) The perverse voyeurism to which emerging technologies give rise.

As Singer (2009a) reminds us, "Immanuel Kant's *Perpetual Peace* (1795) first expressed the idea that democracies are superior to all other forms of government because they are inherently more peaceful and less aggressive. This 'democratic peace' argument (cited by presidents across the partisan spectrum from Bill Clinton to George W. Bush) is founded on the belief that democracies have a built-in connection between their foreign policy and domestic politics that other systems of government lack. When the people share a voice in any decision, including whether to go to war, they are supposed to choose more wisely than an unchecked king or potentate."

In other words, since we know that war can bring both victory and glory, or death and despair, and since it directly affects citizens and all their loved ones, democracies strongly pressurize their leaders and urge them to caution and to responsibility rather than to irresponsible adventures. Indeed, glory is mostly to the benefit of the leader, while the loss of loved ones befalls the ordinary citizen. Not forgetting that, in past wars, citizens who had stayed at home, even if they had no friends or relatives at the front, had to face rationing of certain products of consumption (food, clothing, gas) or pay a war tax to sustain the war effort.

But what happens if one sends mercenaries and robots to war instead of citizens, and that one has to put up with neither taxes nor rationing? There will be a general disinterest in the war. One is reminded of it only an instant at the airport when one's toothpaste is confiscated because it exceeds the 100 ml limit. In any case, the influence of public opinion in democratic nations is more theoretical than a reality. The United States of America have fought on many fronts in the last half century, from Korea to Vietnam, from the Persian Gulf to Yugoslavia, from Afghanistan to Iraq, not counting all the minor interventions in Latin American nations. However, the last formal declaration of war goes back to 1941. Italy as well has circumvented the constitutional obstacle that only allows for defensive war and classified foreign interventions (the Gulf War, the attack on Yugoslavia, the invasions of Afghanistan and of Iraq, the intervention in Lebanon, etc.) as 'international police operations' or as 'humanitarian interventions.'

The argument put forward by Korb, Singer, and other experts in robotic weaponry is therefore the following: if 21st century wars no longer require the approval of Congress, if there is no rationing, if no special taxes are imposed, and last but not least machines are made to fight instead of humans, then political leaders will be ever more at liberty and have ever better reasons to opt for military interventions.

To give just one example, faced with the massacres in some of the African nations that we have recently observed (think of the ethnic cleansing done in Rwanda, with children and grown ups actually beaten to death with machetes), Western nations have felt impotent. It could have been politically risky to send troops (perhaps via conscription) into such tough conditions, even with the good intention to save children and innocents. Massive losses would lead to electoral defeat of those politicians taking that decision. But if we had the 'robotic weapons of the future,' the decision might have been another. Predators and Reapers, controlled from Nevada or a European base, could have massacred the Rwandese irregular military bands, and saved the lives of many unarmed civilians, without jeopardising the lives of compatriot soldiers. Therefore this is an attractive argument that it will be ever harder to resist, both for the government and for public opinion.

The second issue Korb raises is that of technological voyeurism. Today Predators see the enemy and kills it. They do exactly what humans used to do at the front. The difference is that human soldiers stored these cruel images inside their brains, that is, in hardware that does not allow file sharing (for the moment being at least). They could tell of what they had seen in their war diaries, or on the radio, or on television. But no one else could see it like they had. Today thousands of movie clips made by drones end online, especially on Youtube, visible, downloadable and distributable by anyone. The military calls these video clips 'war porn' because they show all the cruelty of war with no censorship. People – also because in fiction films they are constantly exposed to violence and phenomena such as spurting blood and exploding brains – are not particularly impressed with death on live. As an example, Singer refers to a video in which a Predator strikes a group of insurgents, having their bodies bounce into the air, while one hears the tune of a pop song by Sugar Ray with the title "I Just Want To Fly." This way war is almost transformed into a sport event, a show, in which the audience is ethically numb, cruel, hungry for revenge, and wants entertainment, and feels none of the compassion that one would expect.

This also happens because the US authorities filter the images and only let through those that serve propaganda. The images that show American soldiers hit, mutilated or killed by the enemy are censored. It would be hard to watch a friend, a son or just someone one knows bounce in the air and pass from website to website, in order to satisfy this kind of pornographic voyeurism. Relatives or friends would have the clip removed. Besides, psychologically, it could have all kinds of effects and unpredictable responses: on the one hand it could increase the desire for revenge, on the other hand it might convince public opinion that war is a pointless bloodshed (that of friends or of the enemy).

War reduced to a videogame, with appropriate filters, could act favourably on public opinion and the ruling classes. Thus, paradoxically, the development of robotic weapons, through decreasing the cost of war in human lives and stress, could in the future increase the number of conflicts as whole, and so increase the level of existential risk for all humanity.

But this is not the only ethical problem. In Singer's (2009a) words, "such wars without costs could even undermine the morality of 'good' wars." A nation's decision to enter war, in order to assist another country that has been aggressed and is close to succumbing, is a moral act especially because that nation is not directly threatened. The moral act lies in the disinterested risk that it takes to lose lives and money. The moral act lies in the collective choice and in the price paid. But if both choice and losses vanished, where is the moral act? "Even if the nation sending in robots in a just war, such as stopping genocide, war without risk or sacrifice becomes merely an act of somewhat selfish charity [...] The only message of a 'moral character' a nation transmits is that it alone gets the right to stop bad things, but only at the time and place of its choosing, and most important, only if the costs are low enough" (Ibid.).

5.5 Analyses and propositions

We have included enough arguments of the ethical kind, for or against the use of robotic weapons. We shall now examine them in the light of the principles and the ethical code that we formulated in the previous chapter.

5.5.1 The Impracticability of the Moratorium

One has, first of all, proposed to bring the robotic arms race to a halt via a moratorium or a ban. Professor Noel Sharkey has formulated the question in precautionary terms, saying in essence that we should hesitate to produce these weapons because they might fall into enemy hands. But this assumes, as its starting point, that only the West is implicated in the manufacturing of these weapons and that hence it is enough to address the editors of the *Roadmap* and a few others to forestall the peril. In reality many nations have for decades been working on robotic weapons systems. As we have seen, drones have already been used, in the 1940s, by the Americans and in the Yom Kippur War by the Israelites, and in addition the Hezbollah in Lebanon and the Pakistanis also have them. It is hard to believe that the Russians and the Chinese have renounced them. It is necessary to understand that there is more than one player and, consequently, no matter how sensible the arguments of the robo-sceptics are, we find ourselves in a classical strategic dilemma, which makes it impossible to make a just choice, for structural reasons that are independent of any single will.

The model is in fact similar to the so-called 'prisoner's dilemma,' an indispensable case-study in every textbook of practical ethics, as well as the basic problem in game theory, which demonstrates how two people (or parties, armies, nations, etc.) might not cooperate, even when cooperation would be in the interest of both.[35] One example of the prisoner's dilemma is the following. Two criminals are arrested by the police. The police does not have sufficient evidence to incriminate them, so it separates the prisoners and visits both of them and gives them the same deal: if the one witnesses in favour of the incrimination of the other (that is, if he defects) and the other remains silent (that is, cooperates), then the accuser is freed and the silent accomplice gets ten years. If both remain silent, both prisoners get just six months in jail for a minor offence. If both betray the other, each is condemned to five years incarceration. Each prisoner must choose whether to betray her/his accomplice or keep quiet. Each is assured that the other prisoner will not be informed that (s)he has been betrayed before the end of the investigation. How should the prisoners act?

Various philosophers and mathematicians have tackled this problem, among whom John Nash, who formulated a solution known as 'Nash's Equilibrium.' One generally agrees on the most likely result of the negotiation. If one assumes that

---

35 Originally elaborated by Merrill Flood and Melvin Dresher at the RAND in 1950, the prisoner's dilemma was later formalized and given its present name by Albert W. Tucker (Poundstone 1992).

all each player wants is to minimize his own time in jail, it follows that the prisoner's dilemma does not form a zero-sum game, in which each player can either cooperate with the other player, or betray her/him. The only equilibrium in this game is a so-called 'Pareto suboptimal' solution, in which the rational choice induces the two players to defect, and get five years, even if the gain for each player would be superior if they cooperated (for just six months).

This dilemma had much success, partially because it was formulated during the Cold War and appeared as a perfect description of the arms race between the USA and the USSR (the two prisoners). It was in the interest of both to stop the race, but mutual lack of confidence impeded cooperation. Nothing has changed much with the robotic arms race, with the difference that now the prisoners are not just two, but many. This renders the solution to the problem at the mathematical level even more complicated.

This is not to say that it would be naïve or useless to state the problem, but simply that it would be naïve to believe that there is an easy solution for it, or that the ethical problem is just a dilemma with a binary choice. To think that one can stop the robotic arms race with a proclamation is like imagining that shouting: "Crimes must cease!" on the rooftops will defeat crime. Crimes can be defeated only if one removes the causes that generate them and at the same time makes sure that the victims and the suspected criminals are not denied their rights. The same goes for robotic weapons. As long as there are wars, nations involved will always want to have the most powerful and sophisticated weapons. So, if these frighten, then one needs to envisage creating a balance of geopolitical forces that makes resorting to war rare and inconvenient. Crying wolf is not enough. We need (and this is a lot harder) to find the lair and tame it.

If convincing a nation to renounce making robotic weapons may seem all but impossible (the reply will be: "Go convince Russia and China, and then we'll talk about it"), the idea however of opening up a debate to regulate its use, also in wartime, is not futile. The same goes for chemical, bacteriological and nuclear weapons. To conclude, one may accept the idea of not using them, but not the idea of not having them.

The goal of 'owning with inhibited use' is perfectly in line with our principles of rational ethics. And it is also compatible with Immanuel Kant's approach to ethics, as well as with some of the principles of ancient traditional morality – Eastern and Western. Effectively Kant's meta-norm known as the 'categorical imperative' can be formulated as follows: "Act in such a way that the maxim of your (subjective) action could become a universal (objective) law." In spirit, if not in letter, it comes close to the principle of reciprocity of Confucian tradition (embedded also in the Gospels): "Do not do to others what you would not want done to you." Applying the categorical imperative (or the principle of reciprocity) to one's own actions, anyone can see if these are moral or not. Thus one could ask a thief: "Would you want burglary to become a universal law, that is, that everybody would steal instead of doing honest work?" It is obvious that *rationally* the thief would have to give a negative reply, because if everybody stole there would be nothing to steal. If rendered universal, the immoral act becomes impossible.

Of course in war the principle of reciprocity has been often violated. In addition, also at the theoretical level, everybody does not accept the idea that ethics must have a rational foundation or be founded on an egalitarian principle such as the one just outlined. Those who view themselves as 'the elect people' or 'a superior race' or 'a nation with a manifest destiny' could give themselves rights and prerogatives that they do not concede to others. But what we want to stress here is that, contrary to what one might think, an egalitarian approach to ethics does not at all rule out belligerent action. The categorical imperative is meaningful also in the context of war and is compatible also with military operations. We will give just one example. We can kill our enemies in a gunfight, with conventional weapons, also because we have accepted the possibility of dying in such a context. However at the same time we can refuse to pluck out our enemies' eyes, because we would never want this to become a universal law and that our own eyes were plucked out, should we become prisoners. In the end, the purpose of a rational approach to ethics is that of creating conventions and rules that are widely shared, also in situations of lethal conflict. And history demonstrates that this is not a chimerical approach. Even during World War II, which, by virtue of its use of devastating weapons and the total number of casualties, has been the most bloodthirsty conflict in human history, none of the fighting powers – however radical the ideological confrontation – violated certain conventions and rules that had been agreed upon: for instance the prohibition to use nerve gas on the battlefield.

To sum up, because of the prisoner's dilemma, it makes little sense to require that nations forgo robotic weapons, especially now that we find ourselves in a phase of history with many and widespread conflicts, but that because of Kant's principle of the categorical imperative, as shown by various historical cases, it becomes possible (and also cautious) to arrive at an international convention that regulates the use of these weapons.

5.5.2 Pragmatism as a remedy for undesired effects

The second major issue Sharkey, Brown, Singer and many others raise has to do with robot errors, to their hypothetical going awry, to the problem of defining the responsibility in the case of slaughter of innocents (as in the emblematic case of the Iranian Airbus in 1988). This is a serious problem with no easy solution, which has occupied both commissions and magistrates. If it is not possible to punish the robot, then it is clear that the responsibilities can be shared (according to the case) among designers, makers and users, as happens with other technological objects.

However let us make one thing clear: the hypothetical elimination of electronic equipment and automatic systems from airplanes, warships and battle tanks does not at all shelter us from possible errors. Human beings are also prone to errors and, worse, deliberate cruelty. When one repeats like a mantra that "the control must remain in human hands," in order to reassure public opinion, this one should ask itself which human hands will indeed control the weapons. Robots may kill civilians by mistake, which indeed is awful, but let us not forget that humans have systematically and deliberately killed civilians out of revenge or cruelty. Think only of indiscriminate bombing of cities in order to sap enemy resistance.

The robot soldier might mistakenly point his weapon at a civilian or kill an enemy that has already surrendered, but the human soldier is capable of worse. He has tortured prisoners, humiliated, mutilated and killed them for the sheer pleasure of it. We can mention the Turks who impaled prisoners, or the Phoenicians who during the Third Punic war mutilated Romans on the walls of Carthage and threw their remains into the air. Or had them crushed by elephants or by the keels of their ships. But without going back this far, it is enough to think of the tortures some US soldiers inflicted on Iraqi prisoners.

Finally, it maybe fruitful to discuss the possibility that robots change (blindly) into potential assassins, but we do not think that these problems could be resolved by simply handing control over to humans. Humans are not angels. They could commit atrocities that are much worse than machine errors. Add to this the fact that technology keeps improving, while humans evolve much more slowly, and the argument from error might be overcome in a couple of decades. In other words, one should not think of control as the negation of all autonomy, but rather as the capacity to stop the machine from functioning should the situation degenerate dramatically.

To put it in even clearer terms, the ethical and cautionary problem, once one has adopted a pragmatic perspective, is not resolved by imposing human control as a matter of principle, but by continuous assessment (and so the old procedure of trial and error), which is the procedure that would offer the best results, that is, to achieve our ends with the fewest casualties, both friendly and enemy. This goal can be obtained with human control, with machine control, or with mixed control. Only experience, and statistics, will tell.

5.5.3 Voyeurism as an antidote to conflict escalation

Let us now take a look at the other issues Singer raises relative to the undesirable effects of robotic war: the trivialization of combat, the probable increase of conflicts, a sick voyeurism of our information society, weakening democracies, a growing gap between civilian society and the military apparatus. These issues are all connected and they are not illusory. Were we certain that political leaders of the future would use robotic arms to halt situations of gross injustice, violence, human rights violations, we would have nothing to fear. When armed militia or the regular army oppress unarmed civilians, children, minorities, then it is likely that intervening political leaders would have the support of public opinion. However, history tells us that political leaders have started wars for much less noble reasons, such as distracting public opinion from internal political problems, or to favour the conquest of new markets on behalf of the economic and financial lobbies that support them – with fake casus belli constructed using mass media controlled by those same lobbies. If one considers that the lack of morality (understood as acting in one's own interest with no respect for others' life and freedom) can nestle also inside the political and economical classes, the alarm – called by the military and civilians interviewed by Singer – seems more understandable. I would worry more about these aspects of the decision process than about the 'weak morality' inherent in a costless military intervention.

As regards the 'porn war,' I think that there is nothing new under the sun. The medium changes (and this is not without its importance), but surely one cannot blame this phenomenon on computers and robots. Think only about Roman gladiators, about the propaganda spread by belligerent nations during the two world wars, and during the cold war, to portray the enemies as inhuman beings who deserve no mercy and one's own weapons as invincible. Of course there are some new psychological elements, but once again we should take a look at human nature rather than trying to solve the problem by banning Predators or footage online.

The porn war that is all the rage on YouTube satisfies a desire for revenge that is inherently human and atavistic. As for the war in the Middle East, it has been fuelled also by insurgents slitting the throat of American prisoners; these have then been picked up and spread online. In other words, the new media have not at all created these instincts from scratch, but they make them visible. It should also be stressed that, while one part of users seem insensitive or even thrilled by looking at such scenes of violence, there have also been reactions of indignation. Therefore these clips – precisely because of their cruel and violent nature – could have also a positive function, because they show public opinion what war really is. By sensitising public opinion, 'war porn' could induce it to take a greater interest in government decisions, and act as a counterweight to the tendency of military interventions to escalate.

5.5.4 Correct information as a counterweight to alarmism

The new robot prototypes under study, especially those who 'feed' on biomass – the EATR model – have also unleashed ethical discussions. On Fox the news were given an alarmist title: "The Pentagon is working on war robots that feed on dead bodies" (R. Z. 2009). This is false. With famous concern, Robot Technology Inc. and Cyclone – the two companies involved in the project – immediately denied this statement, and clarified that theirs is a vegetarian robot. But despite the clarification the press insisted. In fact, Italian press agency *AdnKronos* (2009) reissued both theses, and this with a hyperbolic title: "Here comes EATR, the war robot that can feed on flesh: the debate on cyberethics heats up." The agency's first bulletin is telegraphic: "Miami – (IGN) – On the battlefield fallen fighters' dead bodies could be the easiest fuel to use for this new robotic weapon that, in order to work, uses biomass. The manufacturing companies deny this: it is 'vegetarian.' But the discussion on the limits to set on these machines, that scientists foresee will soon be able to make autonomous choices, in order to avoid ethical conflicts has been going on for some time."

Later a more detailed revision, but alarming nonetheless, was published on the agency's website: "Miami, Aug. 21st, (IGN) – The robots to which we are used today are at best copying dogs or act as vacuum cleaners that run about the house looking for tiny scraps. But, ever faster, 'mechanical creatures' tackle complex tasks, perhaps on the battle stage, as is happening in Iraq and Afghanistan. The latest robot soldier to arrive on the scene, a transport vehicle that moves fuelled by a biomass engine, that is, it burns organic stuff to run, generates some hesitation in the cybernetic world. Indeed, on the battlefield the most common fuel available might well be human flesh" (Ibid.).

The problem of robots eating human flesh is on the desk, even though it is now spoken of as a merely academic hypothesis. After all, it is true that there are dead bodies on the battlefield and it is true that the robots feed on biomass, and since dead bodies are biomass, if one plus one equals two... the rest is consequence.

But perhaps this idea arose in someone's head because of its name? "It is called EATR – which in English sounds uncannily like 'eater.'" And yet the makers cannot be clearer. Harry Shoell, manager at Cyclone, puts it thus: "We completely understand the public's concern about futuristic robots feeding on the human population, but that is not our mission," and he adds that no one would dream of violating article 15 of the "Geneva convention" that prohibits the desecration of the corpses of the fallen. The reporter has to take note of this: "The engine developed to power the EATR runs on fuel no scarier than twigs, grass clippings and wood chips."

Yet, can one easily disregard so gluttonous a piece of news? The humanoid cannibal makes splashier headlines than a lawnmower that recycles greens, so it is better to stress the academic hypothesis: "What would happen, critics ask, if it malfunctioned or ran out of fuel? It would make do with whatever it found, is the answer, and conceive of worrying scenarios along the lines of 'Terminator' or 'Matrix,' science fiction movies where machines take over the planet and use humans as a source of energy" (Ibid.).

Even though the news is really farfetched, the reporter is right to raise the ethical problem:

> In cybernetics the problem of what ethical boundaries should be imposed on these mechanical creations does not go away, given that scientists foresee that very soon it will be possible to make robots able to make largely autonomous decisions. The science fiction writer Isaac Asimov, the author of *I, robot*, had for this purpose conceived three simple laws which, in a remote future, would be programmed into the electronic brains of the automatons. The first of the three, fundamental this one, states: 'A robot may not injure a human being or, through inaction, allow a human being to come to harm.' But he certainly had not taken into consideration the problem of a robot which, in order to exist, might be forced to eat human flesh (Ibid.).

Once again, we are not so worried about the actual performance of the machine or the use that one will make of it (it has not yet been used), but the fact that it violates a certain idea of how the world works, to put it like Brown. Ordinary people are convinced that there is a neat, ontological, separation between the animal reign and the vegetal reign, the organic and the inorganic, the living and the dead, the conscious and the unconscious. Robots and GMOs demonstrate that these distinctions are just a convenient heuristic model to classify objects, while reality is much more complex and plastic. A robot can draw energy from his environment and feed himself no more no less like a human being or an animal. With the addition that if there be no potatoes or carrots it can also run on gas or petrol. This worries people because it appears to cast into doubt the uniqueness of humans.

Moreover, the mere existence of EATR conveys that it is at least *technically possible* to build a robot that kills humans and feeds on their flesh, so that it could run for an undetermined length of time. To stop it one would have to switch it off (put it to sleep) or destroy it. If there is no such model it is only because *Robotic Technology* decided to make it vegetarian. Human creative power fascinates some people and frightens others. Hence the ethical controversy. From a pragmatic and rational point of view, it is advisable to serenely accept the 'fact' that the boundaries between organic and inorganic are transient, and strive for these machines to generate more happiness than unhappiness in the world. Taking it for granted that they don't have feelings (happiness or despair), it would be suitable to give priority to humans and therefore give them the authority to stop the machines at any time in case of malfunctioning or unforeseen and negative collateral effects. However, it seems rational to try also to take advantage of it for civilian or military use. After all, EATR is the ideal lawnmower both as to performance and to save energy. And it would be the only robot able to hinder the action of enemy soldiers or militia over many days in a hostile environment, far from the bases and cut off the system of logistic assistance.

5.6 Scenario analysis: dreams and nightmares

What will happen tomorrow? If humans rationally tended to choose what is 'good' and to reject what is 'bad,' for themselves and for others, in theory we ought to see a constant improvement of the human condition. But this can only happen in utopias. The problem is that human choices are not always free. They are not always rational. What is good to one group is not always good for another. What is rational at the micro level (individual, social group) is not always rational at the macro level (society, humanity), and vice versa. And finally there is always the possibility of the 'unanticipated consequences of purposive actions,' already studied in detail by sociologist Robert K. Merton (1936). That is, even if we assume social actors to be rational and have positive intentions, there can always be undesired collateral effects. As a popular saying goes: "The road to hell is paved with good intentions."

For the time being, the development of robotics appears unstoppable. We keep hearing that the 21st century will be the century of robots. This happens because, on the whole, such a development appears 'good,' despite the above-mentioned worries and concerns. It appears 'good' also because the classical idea of virtue as a capacity (courage, knowledge, rationality, self-discipline, ability) is once more in favour, and there is no doubt that robots are 'good' in this specific sense. And their 'parents' are every bit as good, since they have been able to transmit to the robots the capacity do to many things. Among these, the ability to fight.

The reason why military applications are being continuously developed is precisely this one: they are 'good soldiers.' First of all they save lives. At the same time they do not have the typically human phobias and weaknesses. In the words of Gordon Johnson of the Pentagon's Joint Forces Command: "They don't get hungry. They are not afraid. They don't care if the guy next to them has just been shot. Will they do a better job than humans? Yes" (Weiner 2005). Add to this that robots, unlike humans, can be trained and can transmit abilities from the one to the other in an extremely short time: download time. This too is a crucial feature, not just for war, but also in the ever more stringent economical conditions.

At the time of the invasion of Iraq, in 2003, only a handful of drones were in use by the V Corps, the primary command force of the US army. "Today – Singer (2009a) writes five years later – there are more than 5,300 drones in the US military's total inventory and not a mission happens without them." Therefore, moving on to predictions, one lieutenant of the US Air Force states that "given the growth trends, it is not unreasonable to postulate future conflicts involving tens of thousands" (Ibid.).

Between 2002 and 2008, the US defence budget grew 74% to reach 515 billion dollars, not counting some hundred billion dollars, spent on the interventions in Afghanistan and Iraq. Within this expense, the investment into making land unmanned systems is to double every year as of 2001. The Pentagon's order to the constructors is unambiguous: "Make them as fast as possible."

Singer again compares the current situation with that of the industrial take-off, shortly before World War I. In 1908 239 T-Ford cars were sold. Ten years later over a million had been sold. We add that similar situations have been observed with the radio, televisions, computers and telephones. When the home robot boom will take place it will be no less sudden than the technological booms preceding it. The presence of these intelligent machines in homes and in the street will astonish at first, and then be taken for granted.

As regards war machines, one has reached a limit in the development of some manned systems, in particular as regards supersonic aircraft. For example, the intercepting fighter F 16 is too good, in the sense that it is a lot better than the human pilots flying it. It can operate at high speed and follow trajectories, which to a human pilot would be beyond the physically and sensorially endurable. Only a properly programmed computer could maximally exploit the mechanical and aerodynamic features of the latest generation supersonic fighters.

This also goes for other weapons systems. If land robots were able to respond to gunfire, by means of laser sensors and pointers to identify the target, we would see extremely quick responses. Assuming that, in the future, armies on the field will also include robotic soldiers, with the gradual shortening of the loop, then it becomes clear that presence of humans will no longer be possible on the battle field: our reaction times are far too slow.

Therefore humans must inevitably be replaced by robots if the possibilities offered by engineering are to be fully exploited. Bluntly, in the words of one DARPA official, we will have to take into account that "the human being is about to become the weak link in the defence system."

This is why the US are getting ready to set up a "Future Combat System" (FCS), at a total cost of 230 billion dollars, that Robert Finkelstein describes as "the largest weapons procurement in history...at least in this part of the galaxy" (Singer 2009b, 114). The basic idea is to replace tens of thousands of war vehicles with new integrated systems, manned and unmanned, and to write a 34 million lines long software program for a network of computers that will connect all the war machines on land and in the air. Each individual brigade will have more land robots on the field than traditional vehicles, with a ratio of 330 to 300, and one hundred drones under the direct control of ground vehicles. The new robotized brigades could be ready for action in the near future.

Future drones will not necessarily resemble Predator or Reaper. We have already hinted at the futuristic shape of engineering's latest gem, the Northrop Grumman X-47, more resembling a fighter in Battlestar Galactica than a traditional airplane. But also giant drones are under construction. They have a wing span the size of football fields, running on solar panels or hydrogen, capable of being in the air for weeks on end, akin to orbiting spies, but easier to operate. Another direction where research is heading is that of miniaturization, or if we want to use a word more in vogue, that of nanotechnology. In 2006 DARPA gave the green light to a research project with the aim to build a drone with the dimensions and performances of an insect, that is, weighing less than 10 grams, being shorter than 7,5 centimetres, capable of flying at 10 meters/second, with a range of action of one kilometre, and able to hover in the air for at least one minute.

A drone of this kind, other than its military uses, could also be used by the secret services and by the police, to spy or kill. Indeed it could function like smart bombs on a smaller scale. The microdrone would revolutionize all the systems of protection and would have no small consequences on politics and on society. Keep in mind that, in the near future, just as it could be in the hands of the police or the army, the mafia and terrorist groups could have it too. If today it is rather hard and risky for terrorists and mafias to try to kill a politician or some other eminent personality, with the aid of these microscopic flying robots it could become all too easy. It would be enough to remote control them with a SAT-NAV system to reach the victim. The microdrone could thus blow up near the head or other vital organs, or even, alternatively, kill the victim with a lethal injection or with a high voltage electric charge, and then fly off. If an almost invisible nanodrone were to be made, it could enter the nostrils or ears of the victim, killing it with a micro-explosion inside the skull, eluding and confusing the traditional system of protection. Indeed it would not be easy to identify the source, unless one had even more sophisticated electronic systems to monitor and intercept. Setting out to build ever more sophisticated systems of protection, that is, antidotes to nanotechnological weapons, seems therefore more important than putting the weapon itself on the market.

In a hitherto unprecedented situation of vulnerability, it could become all but unsuitable to have a public role in politics, media or entertainment – particularly if such a role is hostile to major powers, mafia or groups with a strong ideological identity. But keep in mind that any 'enlightened lunatic' – laying his hands on this kind of weapons system – could try to kill a famous or powerful person out of sheer envy or paranoia. Probably, other than systems ID, it would be fitting to prepare a rather rigorous system of traceability that will include satellite systems and systems of land spies able to intercept almost any nanodrone or microdrone in the air or on the ground.

Excessive alarmism could be premature or unfounded, because in history every weapon has had a shield able to stop it. When we went online for the first time and our computers were aggressed by the first viruses, some said that the Internet would never take off as a tool for the masses, because the very expensive hardware could be systematically destroyed by virulent software costing next to nothing. One had not taken antiviruses into account and one had not taken into account the fact that some software would have cost more than the hardware themselves. Of course, more than one user had his computer destroyed by a virus. But these annoying incidents have not taken down the system.

This is to say that the predictions that we are venturing here can only be pure speculation. The future that nanotechnology will generate cannot be foreseen in full. In 2007, when David Leigh, a researcher at the University of Edinburgh, managed to construct a 'nanomachine' the individual parts of which were of the dimension of a molecule, we understood that technology had suddenly projected us into a novel direction with unpredictable consequences. If historical eras are defined by materials (stone, copper, bronze, iron, plastic, etc.), then we have entered into the age of nanomaterials (Serreli, Lee, Kay, and Leigh 2007). What will it bring us? Leigh could not tell: "It is a bit like when stone-age man made his wheel, asking him to predict the motorway" (as cited in Singer 2009b). We have entered into a new world, but it is simply impossible to know which kind of world it will be. Any presumption to do so will therefore miss the mark.

The future will be a world of nanomachines, but also the world of androids. An android (or a humanoid) is a robot resembling a human and able to imitate many human behaviours; many designers hope that they will also be able to think and feel in ways analogous – even though not absolutely identical – to those of humans. Ian Pearson had defined 'androids' as machines that have a consciousness, linking the concept not so much to the anthropoid shape, as to the anthropoid mind. Scientists and engineers are already designing humanoid soldiers (Nath and Levinson 2014).

The military hopes that androids – whatever is meant by them – will be even better warriors than humans. When DARPA asked the military and scientists to indicate what role robots will play alongside humans, and then without them, in the near future, they replied in the following order: demining, recognisance, vanguard, logistic, and infantry. Oddly, air defence and driving vehicles, where their use is common, were mentioned only at the end. When they were asked to give a date when it will be possible to send humanoid robots to the battlefield instead of infantrymen, the military said 2025 and the scientists 2020. Robert Finkelstein, president of Robotic Technology Inc., finds these forecasts too optimistic and gives 2035 as the date when androids will first be sent to the front. In any case it is not a long time. Many readers of this book will still be among us to verify the prediction.

5.7 Conclusions

Since the world began, wars have been fought by 'mixed' armies under various flags: an alliance of humans, animals and machines on the one hand, against an alliance of humans, animals and machines on the other. This was the case in the days of Alexander the Great and it is the case today. The war machines that Archimedes or other Hellenistic engineers conceived are not as powerful as the robots we today send out to the battlefield, but still they are their cultural ancestors (Campa 2010a). To wonder if the Golem model will arrive is like asking: will this pattern change? The ongoing emergence of sophisticated objects that violate ordinary people's expectations as to how the world works or should work leads one to suspect that war as a whole could also yield some surprises. The greatest fear is that of seeing, for the first time in history, homogenous and no longer mixed deployments, namely: machines against humans. Science fiction and apocalyptic journalism insist on this matter.

Any prediction, even when founded on rigorous studies of trends, always have a large speculative component by virtue of the complexity of the system. All the same, scenario analyses are still useful and therefore we will not shy away from venturing a prediction. All our analyses lead to the conclusion that the hypothesis of a 'species' war between humans and machines, ending with the defeat of the former, is highly unlikely in the 21st century. The reasons underlying this belief are all in all six.

1) Metaphysical Uncertainty. One must consider first of all that it might be impossible for human consciousness to understand itself or replicate by scientific means. Even though materialistic metaphysics has shown itself most fecund

to science in the last few centuries, and thereby made a privileged hypothesis, this does not allow us to exclude with absolute certainty the plausibility of idealistic or dualistic metaphysics. If the supporters of dualistic mind-matter ontology – like Pythagoras, Plato, René Descartes, Karl Popper, etc. – are correct, then robots can never be conscious in the same way as a human being.

2) The Complexity of Consciousness. Even if we postulate that materialistic metaphysics is correct, it is necessary to acknowledge how *hard* our task is. There has been remarkable progress in Logic, Computer Science, Psychiatry, Biology and Philosophy of Mind in the last centuries, but we are still a long way from understanding the concept of consciousness. And we cannot replicate what we do not understand. We can only make something different. In addition, considering that we have not yet managed to solve technical problems that are apparently simpler, such as a cure for baldness or caries, it is understandable that some previsions about the technological development of androids are regarded as overly optimistic.

3) The Alien Character of Artificial Consciousness. Even if we postulate that consciousness is just an emerging property of matter when suitably organized, and admit that artificial consciousness could emerge as an undesirable collateral effect from other actions, this does not imply that alien intelligence would necessarily be a hostile artificial intelligence. In other words, even if our machines were to spontaneously acquire their autonomy for reasons beyond our comprehension, this does not logically entail that they will be violent towards us. We tend to view humans as angels and machines as potential Terminators, but all anthropological and biological observations demonstrate that it is man in fact who is the most dangerous and aggressive predator produced by evolution. An alien intelligence could be benevolent precisely because it is alien, and not in spite of it. In other words, the alien character of artificial intelligence is in reality an argument against it being hostile. This is how things stand now until proven otherwise.

4) Potency of Technological Man. Even if a hostile artificial intelligence were to emerge, even if our robots were to rebel against us, humans are still powerful enough to engage in the equivalent of an 'ethnic cleansing' of the machines. Let us not forget that humans would not be fighting the robots with bows and arrows, but with blinded tanks, airplanes, remote controlled missiles, and, in extreme cases, nuclear devices. The battle would be between two hitherto unseen armies: on the one hand an alliance of manned systems and unmanned systems that have remained faithful to humans, and on the other hand unmanned systems remote controlled by hostile artificial intelligence. The final outcome of this hypothetical clash is anything but certain.

5) Evolution of Technological Man. Even if unmanned systems were to evolve to the point of becoming more potent than any manned system, we should not forget that humans themselves will presumably undergo an evolution by technological means. Humans, using genetic engineering or hybridising with machines via the implants of microchips in the brain or under the skin, could cease to be the weak link in the chain. In the future they might react at a thinking level equal in speed and precision to those of machines.

6) Man-Machine Hybridization. Finally we must consider that, because of technological development in the fields of bioengineering and of robotic engineering, we might never have a conflict between the organic and the inorganic worlds, between humans and machines, between carbon and silicon, simply because there will be a real and true ontological 'remixing.' There will be human beings empowered with electro-mechanical parts and robots with organic portions in their brain. Therefore it is not ontology that will decide the alliances.

In conclusion, we believe that in the 21st century there will still be humans, machines and animals serving under one national flag, waging war against humans, machines and animals serving under another national flag. When this system has disappeared, if there are still conflicts, in our opinion it will be more likely to see a variety of sentient beings (humans, transhumans, and posthumans) on the one hand, under one flag, against a variety of sentient beings (humans, transhumans, and posthumans), under another flag. But we are speaking of a very remote future.

The more concrete and pragmatic recommendation that I would now give makers of robotic weapons and their political and military customers is to always work on parallel projects, conceiving, for each robotic weapon that they construct, another weapon able to control and destroy it. This precaution could reveal itself useful both in the science fiction scenario of the emergence of hostile artificial intelligence, and in the more prosaic and plausible scenario that the robotic weapon falls into enemy hands.

However I believe it inopportune and irrational to apply the maximalist version of the precautionary principle. By maximalist version I mean an interpretation of 'precaution' that would mean banning any technology that does not present itself as absolutely risk-free.[36] First of all, there is no technology or human action that

---

36 On this problem we invite the reader to have a look at Petroni (2009). Even though he mainly focuses on the biotechnologies, the article offers a detailed and convincing analysis of the precautionary principle.

is risk-free, because it is not possible to foresee the whole range of future developments inherent in a certain choice. As it is said, the flapping of a butterfly wing in the Southern Hemisphere can cause a hurricane in the Northern Hemisphere. Second, since we do not live in paradise and since processes pregnant with a future that we do not know are already in the making, non action does in no way guarantee that the results will be better for our group or for all humanity. This to say that the failure of the butterfly wing to flap in the Southern Hemisphere could also provoke an extremely serious drought in the Northern Hemisphere. Finally, the precautionary principle (at least in its maximalist interpretation) never pays sufficient attention to the benefits that might derive from a risky action. On closer inspection, fire has been a risky undertaking for Homo Erectus. During the million or maybe more years that separate us from the discovery of the technique of lighting and controlling fire, many forests and cities have been consumed by flames because of clumsy errors by our ancestors. Billions of living beings have probably died because of this technology. And today we still hear of buildings that burn or explode, causing deaths, because of malfunctioning central heating systems or mere forgetfulness.

Yet what would humans be without fire? If our ancestors had applied the maximalist precautionary principle, rejecting fire because it is not risk-free, today we would not be Homo Sapiens. This dangerous technology has indeed allowed us to cook our food and hence for hominid jawbone to shrink, with the ensuing development of language, and of the more advanced idea of morality and technology that language allows. In brief, today we would not even argue in favour or against the precautionary principle, or indeed any principle, because these require language for their formulation.

Bibliography

AdnKronos, 2009, August 21. "Creato EATR, robot da guerra che può nutrirsi di carne: s'infiamma il dibattito sulla cyber-etica," *AdnKronos*.

Ahmed, A. 2013. *The Thistle and the Drone. How America's War on Terror Became a Global War on Tribal Islam*. Washington D.C.: Brookings Institution Press.

Aridas, T. and Pasquali, V. 2013, March 7. "Countries with the Highest GDP Average Growth, 2003–2013," *Global Finance*.

Aristotle, 1995 (350 B.C.E.). *Politics*, Books I and II. Oxford: Clarendon Press.

Ayres, R. U. 1998. *Turning Point. The End of the Growth Paradigm*. London: Earthscan Publications.

Bashir, S. and Crews, R. D. (eds.) 2012. *Under the Drones. Modern Lives in the Afghanistan-Pakistan Borderlands*. Cambridge (MA) and London: Harvard University Press.

Bennato, D. 2004, April 19. "Roboetica: un caso emblematico di tecnoetica." www.tecnoetica.it (accessed January 1, 2011).

Bignami, L. 2007, April 10. "Robot, la grande invasione," *La Repubblica*.

Blaug, M. 1958. *Ricardian Economics*. New Haven: Yale University Press.

Blaug, M. 1978. *Economic Theory in Retrospect*. Cambridge: Cambridge University Press.

Bonaccorso, G. 2004. "Roboetica: tra fantascienza e realtà. Analisi critica delle tre leggi di Asimov." www.psicolab.net (accessed January 1, 2011).

Bonaccorso, G. 2011. *Saggi sull'Intelligenza Artificiale e la filosofia della mente*. Morrisville: Lulu.com.

Boyer, R. 2012. "The four fallacies of contemporary austerity policies: the lost Keynesian legacy," *Cambridge Journal of Economics*, 36 (1), 283–312.

Brown, A. 2009, March 18. "War crimes and killer robots," *The Guardian*.

Calvino, I. 1962. *The Nonexistent Knight and The Cloven Viscount*. San Diego: Harcourt Brace Jovanovich.

Campa, R. 2004. "La *Storia filosofica dei secoli futuri* di Ippolito Nievo come caso esemplare di letteratura dell'immaginario sociale: un esercizio di critica sociologica," *Romanica Cracovensia*, 4, 29–42.

Campa, R. 2006. "Transumanesimo," *Mondoperaio*, N. 4–5, July-October, 148–153.

Campa, R. 2007. "Considerazioni sulla terza rivoluzione industriale," *Il Pensiero Economico Moderno*, Anno XXVII, N. 3, July-September.

Campa, R. 2010a. "Le radici pagane della rivoluzione biopolitica," in Id. (ed.), *Divenire. Rassegna di studi interdisciplinari sulla tecnica e il postumano*, vol. 4. Bergamo: Sestante Edizioni.

Campa, R. 2010b. *Mutare o perire. La sfida del transumanesimo*. Bergamo: Sestante Edizioni.

Campa, R. 2011. *Le armi robotizzate del futuro. Intelligenza artificialmente ostile? Il problema etico*. Roma: CEMISS.

Campa, R. 2014a. "Workers and Automata. A Sociological Analysis of the Italian Case," *Journal of Evolution and Technology*, Vol. 24 Issue 1, February, 70–85.

Campa, R. 2014b. "Technological Growth and Unemployment: A Global Scenario Analysis," *Journal of Evolution and Technology*, Vol. 24 Issue 1, February, 86–103.

Čapek, K. 2004 (1921). *R.U.R. (Rossum's Universal Robots)*. London: Penguin Books.

Cappella, F. 2008, March 20. "Big Dog," *Neapolis*.

Carr, A. Z. 1968. "Is Business Bluffing Ethical?," *Harvard Business Review*, January-February, 2–8.

Chomsky, N. and Vltchek, A. 2013. *On Western Terrorism. From Hiroshima to Drone Warfare*. London: Pluto Press.

Clapper, J. R., Young, J. J., Cartwright, J. E., and Grimes, J. G. 2007. *Unmanned Systems Roadmap 2007-2032*, Department of Defense (USA).

Clapper, J. R., Young, J. J., Cartwright, J. E., and Grimes, J. G. 2009. *Unmanned Systems Roadmap 2009-2034*, Department of Defense (USA).

Corriere della sera, 2003, July 29. "Un robot dipinge guidato da un cervello di topo," *Corriere della sera*.

Daerden, F., and Lefeber, D. 2000. "Pneumatic Artificial Muscles: actuators for robotics and automation," Vrije Universiteit Brussel. http://lucy.vub.ac.be (accessed February 27, 2015).

Desai, J. P., Dudek, G., Khatib, O., and Kumar V. (eds.) 2013. *Experimental Robotics*. Heidelberg: Springer.

Devereux, T. 1993. *La guerra elettronica: arma vincente 1812-1992*. Varese: Sugarco Edizioni.

Di Nicola, P. 1998. "Recensione: Luciano Gallino, *Se tre milioni vi sembran pochi. Sui modi per combattere la disoccupazione*." http://www.dinicola.it (accessed April 2, 2015).

Eurostat, 2009. *Science, technology and innovation in Europe*. ec.europa.eu (accessed May 15, 2015).

Evangelista, M. and Shue, H. (eds.) 2014. *The American Way of Bombing. Changing Ethical and Legal Norms, from Flying Fortresses to Drones*. Ithaca and London: Cornell University Press.

Feletig, P. 2010, February 1. "Robot per mare, per cielo e per terra ormai in guerra si va senza uomini," *la Repubblica*.

Fitter, N. T. and Nichols, P. M. 2015. "Applying the Capability Approach to the Ethical Design of Robots." www.openroboethics.org (accessed May 15, 2015).

Floreano, D., and Mattiussi, C. 2008. *Bio-Inspired Artificial Intelligence. Theories, Methods, and Technologies*. Cambridge: The MIT Press.

Ford, M. 2009. *The lights in the Tunnel. Automation, Accelerating Technology and the Economy of the Future*. Acculant Publishing. thelightsinthetunnel.com (accessed April 2, 2015).

G. G. 2008, February 27. "Fermare la corsa alle armi robotiche. Un appello degli esperti di robotica contro il rischio di proliferazione di questi nuovi strumenti di morte," *Le Scienze. Edizione Italiana di Scientific American*.

Gallino, L. 1998. *Se tre millioni vi sembrano pochi. Sui modi per combattere la disoccupazione*. Torino: Einaudi.

Gallino, L. 1999, January 13. "Disoccupazione tecnologica: quanta e quale perdita di posti di lavoro può essere attribuita alle nuove tecnologie informatiche," *MediaMente Biblioteca Digitale*. www.mediamente.rai.it (accessed April 2, 2015).

Gallino, L. 2003. *La scomparsa dell'Italia industriale*. Torino: Einaudi.

Ganapati, P. 2009, July 22. "Robo-Ethicists Want to Revamp Asimov's 3 Laws," *Wired*.

Gangl, M. 2003. *Unemployment Dynamics in the United States and West Germany. Economic Restructuring, Institutions and Labor Market Processes*, Berlin Heidelberg: Springer Verlag.

Ge, S. S., Khatib, O., Cabibihan, J.-J., Simmons, R., Williams, M.-A. (eds.) 2012. *Social Robotics. Fouth International Conference. Proceedings*. Heidelberg: Springer.

Ge, S. S., Li, H., Cabibihan, J.-J., and Tan, Y. K. (eds.) 2010. *Social Robotics. Second International Conference. Proceedings*. Heidelberg: Springer.

Geroni, A. 2011, March 9. "Trenta fallimenti al giorno," *Il Sole 24Ore*.

Giavazzi, F. and Pagano, M. 1990. "Can Severe Fiscal Contractions be Expansionary? Tales of Two Small European Countries," *NBER Macroeconomics Annual*, 5, 75–111.

Giugni, M. (ed.) 2009. *The Politics of Unemployment in Europe. Policy Responses and Collective Action*. Farnham: Ashgate Publishing Limited.

Gordon, R. J. 2004. *Productivity Growth, Inflation, and Unemployment. The Collected Essays of Robert J. Gordon*. Cambridge: Cambridge University Press.

Graeber, D. 2013, August 17. "On the Phenomenon of Bullshit Jobs," *Strike! Magazine*.

Groot, L. 2004. *Basic Income, Unemployment and Compensatory Justice*, New York: Springer Science+Business Media.

Gunkel, D. J. 2012. *The Machine Question. Critical Perspectives on AI, Robots, and Ethics*. Cambridge (MA) and London: The MIT Press.

Habermas, J. 2003. *The Future of Human Nature*. Cambridge: Polity Press.

Habershon, E., and Woods, R. 2006, June 18. "No sex please, robot, just clean the floor," *The Sunday Times*.

Haddadin, S. 2014. *Towards Safe Robots. Approaching Asimov's 1st Law*. Heidelberg: Springer.

Harnad, S. 2003, "Can a Machine Be Conscious? How?," *Journal of Consciousness Studies* 10 (4), 67–75.

Hermann, G., Pearson, M. J., Lenz, A., Bremner, P., Spiers, A., and Leonards U. (eds.) 2012. *Social Robotics. Fifth International Conference. Proceedings*. Heidelberg: Springer.

Heron, 1976 (1899–1914). *Heronis Alexandrini opera quae supersunt omnia*, 5 vol., Leipzig, Teubner.

Horgan, J. 1997. *The end of science. Facing the Limits of Knowledge in the Twilight of the Scientific Age*. New York: Broadway Books.

Howell, D. R. (ed.) 2005. *Fighting Unemployment. The Limits of Free Market Orthodoxy*, Oxford: Oxford University Press.

Hughes, J. (ed.) 2014. *Technological Unemployment and the Basic Income Guarantee*, Special issue of *The Journal of Evolution and Technology*, Volume 24 Issue 1.

Hughes, J. 2004. "Embrace the End of Work. Unless we send humanity on a permanent paid vacation, the future could get very bleak," *USBIG Discussion Paper No. 81*. www.usbig.net (accessed April 2, 2015).

Istat, 2010. *La popolazione straniera residente in Italia al 1 gennaio 2010*, October 12. www.istat.it (accessed April 2, 2015).

Istat, 2011. *Censimento dell'Industria e dei Servizi*. www.istat.it (accessed April 2, 2015).

Kanda, T., and Ishiguru H. 2013. *Human-Robot Interaction in Social Robotics*. Boca Raton: CRC Press.

Keynes, J. M. 1930. "Economic Possibilities for our Grandchildren." In: Idem, *Essays in Persuasion*, New York: W. Norton & Co., 358–373.

Kramer, S. N. 1997. *I Sumeri. Alle radici della storia*, Newton, Roma [English title: *History Begins At Sumer*, University of Pennsylvania Press, 1988.].

Krugman P. 2013, June 13. "Sympathy for the Luddites," *New York Times*.

Kurfess, T. R. (ed.) 2012. *Robotics and Automation Handbook*. CRC Press. Kindle Edition.

Kurz, H. D. 1984. "Ricardo and Lowe on Machinery," *Eastern Economic Journal* Vol. 10 (2), April June, 211–229.

Lin, P., Abney, K. and Bekey, G. A. (eds.) 2012. *Robot Ethics. The Ethical and Social Implications of Robotics*. Cambridge (MA): MIT Press.

Lira, 2015. "Babybot on the Press." http://www.lira.dist.unige.it (accessed April 6, 2015).

Liu, Y., and Sun, D. 2012. *Biological Inspired Robotics*. Boca Raton: CRC Press.

Lowe, A. 1954. "The Classical Theory of Economic Growth," *Social Research*, Vol. 21, 127–158.

Lowe, A. 1976. *The Path of Economic Growth*. Cambridge: Cambridge University Press.

Malinvaud, E. 1994. *Diagnosing unemployment*. Cambridge: Cambridge University Press.

Marshall, G. 2003. *Oxford Dictionary of Sociology*. Oxford: Oxford University Press.

Martorella, C. 2002, October 5. "Shigoto. Lavoro, qualità totale e rivoluzione industriale giapponese." http://cristiano-martorella-archivio.blogspot.com (accessed April 2, 2015).

Maruccia, A. 2010, February 28. "Era del Robot, umanità in pericolo," *Punto Informatico*.

Marx, K. 1976 (1867). *Capital. A Critique of Political Economy*, Vol. 1, Harmondsworth and New York: Penguin Books.

Marx, K. and Engels F. 1948 (1888). *Manifesto of the Communist Party*, New York: International Publishers.

Meggiato, R. 2010, July 16. "EATR il robot vegetariano. Sviluppato il primo robot in grado di alimentarsi da sé con dei vegetali," *Wired*.

Menghini, M., and Travaglia M. L. 2010. *L'evoluzione dell'industria italiana. Peculiarità territoriali*. Istituto Guglielmo Tagliacarne. www.tagliacarne.it (accessed April 2, 2015).

Merton, R. K. 1936. "The unanticipated consequences of purposive social action," *American Sociological Review*, 1, 894–904.

Merton, R. K. 1968. *Social Theory and Social Structure*. New York: Free Press.

Mill, J. S. 1848. *Principles of Political Economy with some of their Applications to Social Philosophy*. www.econlib.org (accessed April 2, 2015).

Minsky, M. 1956. Some Universal Elements for Finite Automata. *Automata Studies: Annals of Mathematics Studies*, Number 34.

Monopoli, A. 2005. *Roboetica. Etica applicata alla robotica*. www.roboetica.it (accessed April 2, 2015).

Monopoli, A. 2007. *Roboetica. Spunti di riflessione*. Morrisville: Lulu.com.

Moravec, H. 1993. "The Age of Robots." www.frc.ri.cmu.edu (accessed April 2, 2015).

Moravec, H. 1993. "The Universal Robot," *Siemens Review*, v. 60/1, January/February, 36–41.

Moravec, H. 1995. "Bodies, Robots, Minds," www.frc.ri.cmu.edu (accessed April 2, 2015).

Moravec, H. 1997. "When will computer hardware match the human brain?," www.transhumanist.com (accessed April 2, 2015).

Mutlu, B., Bartneck, C., Ham, J., Evers, V., and Kanda T. (eds.) 2011. *Social Robotics. Third International Conference. Proceedings*. Heidelberg: Springer.

Nadotti, C. 2006, May 29. "Tokyo prepara le regole per i robot: Non danneggino gli esseri umani," *La Repubblica*.

Nath, V. and Levinson, S. E. 2014. *Autonomous Military Robotics*. Heidelberg: Springer.

New Scientist, 1980. "Car firm drives toward new robot technology," *New Scientist*, June 12, Vol. 86, No. 1205.

Nichols, M. 2013, August 1. "Italian firm to provide surveillance drone for U.N. in Congo," *Reuters*.

Noble, D. F. 1995. *Progress without People. New Technology, Unemployment, and the Message of Resistance*. Toronto: Between the Lines.

Northrop Grumman, "X-47B UCAS Makes Aviation History…Again! Successfully Completes First Ever Autonomous Aerial Refueling Demonstration." www.northropgrumman.com (accessed May 13, 2015).

Operto, F. 2004. "Roboetica da tutto il mondo," *Fondazione Informa*, 6, n. 1. www.fondazionecarige.it (accessed January 1, 2011).

Ostrup, F. 2003. *Money and the Natural Rate of Unemployment*. Cambridge: Cambridge University Press.

Page, L. 2008, April 11. "US war robots in Iraq 'turned guns' on fleshy comrades," *The Register*.

Paprotny, I., and Bergbreiter, S. 2014. *Small-Scale Robotics: From Nano-to-Millimeter-Sized Robotic Systems and Applications*. Heidelberg: Springer.

Pellicani, L. 2007. *Le radici pagane dell'Europa*. Soveria Mannelli: Rubbettino.

Petroni, A. M. 2009. "Liberalismo e progresso biomedico: una visione positiva." In: R. Campa (ed.), *Divenire. Rassegna di studi interdisciplinari sulla tecnica e il postumano, Vol. 2*. Bergamo: Sestante Edizioni, 9–43.

Polchi, V. 2011, March 11. "Il governo ora chiede più immigrati," *La Repubblica*.

Polidori, A. 2004. "Babybot, bimbo artificiale," *Gabinus Rivista Culturale*, II, n. 2. www.lira.dist.unige.it (accessed April 2, 2015).

Potter, S., DeMarse, T. B., Wagenaar, D. A., and Blau, A. W. 2001. "The Neurally Controlled Animat: Biological Brains Acting with Simulated Bodies," *Auton Robots*, 11(3), 305–310.

Poundstone, W. 1992. *Prisoner's Dilemma*. New York: Doubleday.

Punto Informatico, 2006, June 20. "I robot dovranno seguire un codice etico", *Punto Informatico*.

Putnam, H. 1964. "Robots: Machines or Artificially Created Life?," *Journal of Philosophy*, 61 (November), 668–91.

R. Z., 2009, July 17. "Tecnologie inquietanti, il Pentagono studia robot da guerra che si nutrono di cadaveri," *Tiscali Notizie*.

Rampini, F. 2003, May 18. "Il robot col cervello di un topo," *La Repubblica*.

Ranisch, R. and Sorgner, S. L. (eds.) 2014. *Post- and Transhumanism. An Introduction*. Frankfurt am Main: Peter Lang Edition.

Relke, D. M. A. 2006. *Drones, Clones, and Alpha Babes. Retrofitting Star Trek's Humanism, Post-9/11*. Calgary: University of Calgary Press.

Reutter, M. 2001. *A Macroeconomic Model of West German Unemployment. Theory and Evidence*. Heidelberg: Springer.

Ricardo, D. 2004 [1821]. *On the Principles of Political Economy and Taxation. Third edition*. Kitchener: Batoche Books.

Rifkin, J. 1995. *The End of Work. The Decline of the Global Labor Force and the Dawn of the Post-Market Era*. New York: Putnam Publishing Group.

Riotta, G. 2005, October 10. "Il filosofo Fukuyama mette in guardia sui rischi di una ricerca senza limiti: 'No a ingegneria genetica come a fascismo e comunismo,'" *Corriere della sera*.

Rogers, A. and Hill, J. 2014. *Unmanned. Drone Warfare and Global Security*. London: Pluto Press.

Rubart, J. 2007. *The Employment Effects of Technological Change. Heterogenous Labor, Wage Inequality and Unemployment*. Berlin Heidelberg: Springer.

Russo, L. 2004. *The Forgotten Revolution. How Science Was Born in 300 BC and Why It Had to Be Reborn*. Berlin: Springer.

Russo, M., and Pirani, E. 2006. "Dinamica spaziale dell'occupazione dell'industria meccanica in Italia 1951–2001." www.academia.edu (accessed April 2, 2015).

Salvadori, N., and Balducci, R. (eds.) 2005. *Innovation, Unemployment and Policy in the Theories of Growth and Distribution*. Northampton Massachusetts: Edward Elgar Publishing Limited.

Sanchez, M. 2009. "Robots Take Center Stage in U.S. War in Afghanistan," March 23. www.foxnews.com (accessed April 2, 2015).

Schor, J. B. 1993. "Pre-industrial workers had a shorter workweek than today's," http://groups.csail.mit.edu (accessed April 2, 2015).

Schor, J. B. 1993. *The Overworked American: The Unexpected Decline of Leisure*. New York: Basic Books.

Scuola di robotica, 2011. *Roboetica*. www.scuoladirobotica.it (accessed January 1, 2011).

Serreli, V., Lee, C., Kay, E. R., and Leigh, D. 2007. "A molecular information ratchet," *Nature*, February 1, 445, 523–7.

Sharkey, N. 2007, August 18. "Robot wars are a realityArmies want to give the power of life and death to machines without reason or conscience," *The Guardian*.

Sharkey, N. 2008a. "Ground for discrimination. Autonomous Robot Weapons," *RUSI Defence Systems*, October, 86–89.

Sharkey, N. 2008b. "The Ethical Frontiers of Robotics," *Science*, Vol. 322, no. 5909, December 19, 1800–1.

Shin, D., Yeh, X., Narita, T., and Khatib, O. 2013. "Motor vs. Brake: Comparative Studies on Performance and Safety in Hybrid Actuations." In: Desai, J. P., et al. (eds.) 2013. *Experimental Robotics*. Heidelberg: Springer.

Siciliano, B. 2013. "Foreword." In: Desai, J. P., et al. 2013. *Esperimental Robotics*, Heidelberg: Springer.

Siciliano, B., and Khatib, O. (eds.) 2008. *Springer Handbook of Robotics*. New York: Springer.

Singer, P. W. 2009a. "Robots at War: The New Battlefield," *The Wilson Quarterly*, Winter.

Singer, P. W. 2009b. *Wired for War: The Robotics Revolution and Conflict in the 21st Century*. New York: The Penguin Press.

Sloggett, D. 2014. *Drone Warfare. The Development of Unmanned Aerial Conflict*. Barnsley: Pen & Sword Aviation.

Sloterdijk, P. 2013. "Rules for the human park." In: Idem, *You must change your life*. Cambridge: Polity Press.

Sonnenberg, B. 2014. *Dependencies and Mechanisms of Unemployment and Social Involvement*. Wiesbaden: Springer.

Sparrow, R. 2007. "Killer Robots," *Journal of Applied Philosophy*, Volume 24, Issue February, 62–77.

Stockhammer, E. 2004. *The Rise of Unemployment in Europe. A Keynesian Approach*. Northampton Massachusetts: Edward Elgar Publishing Limited.

Sullins, J. P. 2013. "An Ethical Analysis of the Case for Robotic Weapons Arms Control," in Podins, K., Stinissen, J. and Maybaum M. (eds.). 2013. *5th International Conference on Cyber Conflict*. Tallinn: NATO CCD COE Publications.

Tabarrok, A. 2003, December 31. "Productivity and unemployment," *Marginal Revolution*.

Turse, N. 2012. *The Changing Face of Empire. Special Ops, Drones, Spies, Proxy Fighters, Secret Bases, and Cyber Warfare*. New York: Dispatch Books.

Unece, 2004. "Over 50,000 industrial robots in Italy, up 7% over 2002. Italy is Europe's second and the world's fourth largest user of industrial robots." www.unece.org (accessed April 2, 2015).

Unece, 2005. "Worldwide investment in industrial robots up 17% in 2004. In first half of 2005, orders for robots were up another 13%." www.unece.org (accessed April 2, 2015).

University of Sheffield, 2008, February 28. "Killer Military Robots Pose Latest Threat To Humanity, Robotics Expert Warns," *ScienceDaily*.

Veruggio, G. 2007. "La nascita della roboetica," *Leadership medica*, n. 10. www.leadershipmedica.com (accessed January 1, 2011).

Veruggio, G. 2010, October 26. "Roboetica: una nuova etica per una nuova scienza", *Almanacco della scienza – Micromega*, 7.

Veruggio, G., and Operto F. 2008. "Roboethics: Social and Ethical Implications of Robotics." In: B. Siciliano, and O. Khatib (eds.) 2008. *Springer Handbook of Robotics*. Berlin Heidelberg: Springer, 1499–1524.

Veruggio, G., and Operto, F. 2006. "Roboethics: a Bottom-up Interdisciplinary Discourse in the Field of Applied Ethics in Robotics," *International Review of Information Ethics*, Vol. 6 (12), 2–8.

Vroman, W., and Brusentsev, V. 2005. *Unemployment Compensation Throughout the World. A Comparative Analysis*. Kalamazoo Michigan: W.E. Upjohn Institute for Employment Research.

Wang, L., Tan, K. C., and Chew, C. M. 2006. *Evolutionary Robotics: From Algorithms to Implementations*. Singapore: World Scientific Publishing.

Weber, M. 2008. *Max Weber's Complete Writings on Academic and Political Vocation*. New York: Algora Publishing.

Weinberger, S. 2008, April 15. "Armed Robots Still in Iraq, But Grounded," *Wired*.

Weiner, T. 2005, February 16. "New Model Army Soldier Rolls Closer to Battle," *New York Times*.

Werding, M. (ed.) 2006. *Structural Unemployment in Western Europe. Reasons and Remedies*. Cambridge Massachusetts: The MIT Press.

Whittle, R. 2014. *Predator. The Secret Origins of the Drone Revolution*. New York: Henry Holt and Company.

Winnefeld, J. A. and Kendall, F. 2011, *The Unmanned Systems Integrated Roadmap FY 2011–2036*, Department of Defence (USA).

Winnefeld, J. A. and Kendall, F. 2013, *The Unmanned Systems Integrated Roadmap FY 2013–2038*, Department of Defence (USA).

Wittes, B. and Blum, G. 2015. *The Future of Violence. Robots and germs, Hackers and Drones*. New York: Basic Books.

Zaloga, S. J. 2008. *Unmanned Aerial Vehicles. Robotic Air Warfare 2017–2007*. Oxford and New York: Osprey Publishing.

Złotowski, J., Weiss, A., and Tscheligi, M. 2011. "Interaction Scenarios for HRI in Public Space." In: Mutlu, B., et al. (eds.) 2011. *Social Robotics*. Heidelberg: Springer.

Znaniecki, F. 1934. *The Method of Sociology*. New York: Rinehart & Company.

Index of names

A
Abney, K. 77, 155
Albus, J. 70–71
Alexander the Great 146
Alexander, J. 21
Amhed, A. 125
Archimedes 146
Aridas, T. 49, 51
Aristotle 55, 77–78, 84, 151
Arkin, R. 79, 92
Asimov, I. 15–16, 32, 80, 83, 86–89, 92, 98, 141, 151, 153–154
Ayres, R. U. 56–57, 151

B
Babbage, C. 37
Bacon, F. 66, 90, 96
Balducci, R. 15, 158
Bartneck, C. 34, 156
Bashir, S. 125, 151
Bekey, G. A. 77, 155
Beltran, C. 27
Bennato, D. 81–82, 101, 104, 107–108, 151
Bergbreiter, S. 30–31 156
Bignami, L. 40, 47–48, 151
Blau, A. W. 28, 157
Blaug, M. 14, 151
Blum, G. 125, 160
Bonaccorso, G. 81–82, 86–87, 151
Boyer, R. 68, 151
Breckman, W. 21
Bremner, P. 34, 154
Brown, A. 127, 137, 141, 151
Brusentsev, V. 15, 159
Bunge, M. 21
Burawoy, M. 21
Bush, G. W. 129, 132

C
Cabibihan, J.-J. 34, 153
Calvino, I. 106, 151
Campanella, T. 97
Čapek, K. 23, 152
Cappella, F. 119, 152
Carr, A. 106, 152
Cartwright, J. E. 110, 152
Chalmers, D. 99
Chew, C. M. 31, 159
Chomsky, N. 125, 152
Christensen, H. 91, 101
Churchill, W. 128
Ciampi, C. A. 79
Clapper, J. R. 110, 152
Clinton, B. 132
Cobb, C. W. 59
Cohen, E. 129
Columbus, C. 98
Comte, A. 11
Conan Doyle, A. 128
Crews, R. D. 125, 151

D
Daerden, F. 30, 152
Dario, P. 79
DeMarse, T. B. 28, 157
Democritus 75
Desai, J. P. 29, 152, 158
Descartes, R. 147
Devereux, T. 129–130, 152
Di Nicola, P. 46, 152
Dick, P. 102
Douglas, P. H. 59

Dresher, M. 135
Dudek, G. 29, 152
Duffy, B. 79

E
Engels, F. 60, 155
Epstein, R. 130
Evangelista, M. 125, 153
Evers, V. 34, 156

F
Feletig, P. 111-113, 115, 153
Finkelstein, R. 121, 144, 146
Fitter, N. P. 77, 153
Flis, M. 21
Flood, M. 135
Floreano, D. 31, 79, 153
Ford, H. 102
Ford, M. 38, 57, 153
Fries, D. 129
Frysztacki, K. 21
Fukuyama, F. 97, 157

G
Galilei, G. 75
Gallino, L. 41, 43-47, 52, 152-153
Ganapati, P. 77, 153
Gangl, M. 15, 153
Ge Shuzhi, S. 34
Geroni, A. 51, 153
Giavazzi, F. 68, 153
Giddens, A. 21
Giugni, M. 15, 153
Gordon, R. 15, 153
Górniak, J. 21
Graeber, D. 72, 154
Grimes, J. G. 110, 152
Groot, L. 15, 154
Gunkel, D. J. 77, 154

H
Habermas, J. 97, 154

Habershon, J. 89, 91, 93, 154
Haddadin, S. 32, 154
Ham, J. 34, 156
Hammond, M. 21
Harnad, S. 99, 154
Harrod, R. F. 60
Hauer, R. 87
Heron 23, 29, 37, 154
Herrmann, G. 34
Hill, J. 110, 157
Horgan, J. 63, 99, 103, 154
Howell, D. R. 15, 154
Hughes, J. 15, 21, 52, 54, 68, 70, 154

I
Inoue, H. 79
Ishiguru, H. 33, 154

J
Johnson, G. 142

K
Kaczynski, T. 64
Kanda, T. 33-34, 154, 156
Kant, I. 132, 136, 137
Kay, E. R. 145, 158
Kendall, F. 113, 116, 160
Keynes, J. M. 14, 68-69, 72, 151, 154, 159
Khatib, O. 29-30, 34, 39, 152-153, 158-159
Koch, C. 99
Korb, L. J. 132-133
Kowalewicz, M. 21
Kramer, S. N. 91, 154
Krugman, P. 57-58, 155
Krzysztofek, K. 21
Kumar, V. 29, 152
Kurfess, T. R. 39, 155
Kurz, H. D. 14, 155

L
Lamm, C. 21
Lee, C. 145, 158
Lefeber, D. 30, 152
Leibniz, G. W. 75
Leigh, D. 145, 158
Lenz, A. 34, 154
Leonardo Da Vinci 29
Leonards, U. 34, 154
Levinson, S. E. 146, 156
Lin, P. 77, 155
Liu, Y. 31, 155
Lowe, A. 14, 155

M
Malinvaud, E. 15, 155
Malthus, T. R. 14
Manzotti, R. 27
Marconi, G. 29
Marshall, G. 11, 155
Martorella, C. 39, 155
Maruccia, A. 126, 155
Marx, K. 14, 48, 55, 58, 60, 71, 78, 155
Mattiussi, C. 31, 153
Mazur, L. 21
McKenna, T. 59
Meggiato, R. 120–121, 155
Menghini, M. 41, 155
Merton, R. K. 21, 59, 142, 155
Messalina 105
Metta, G. 27
Meyrink, G. 108
Mill, J. S. 56, 155
Milne, A. A. 128
Minsky, M. 37, 156
Monopoli, A. 48, 77, 82, 93–96, 98, 104–105, 107, 156
Moore, G. 25, 52
Moravec, H. 25–28, 37, 52, 62–64, 69, 70, 73, 75, 156
Mutlu, B. 34, 156, 160

N
Nadotti, C. 88–89, 156
Narita, T. 30, 158
Nash, J. 135
Natale, L. 27
Nath, V. 146, 156
Nichols, M. 115, 156
Nichols, P. M. 77, 153
Nietzsche, F. 66
Nievo, I. 102, 151
Nobel, A. 79
Noble, D. F. 55–56, 156

O
Olivero, A. 52
Operto, F. 78–80, 82, 156, 159
Ostrup, F. 15, 156

P
Pagano, M. 68, 153
Page, L. 117, 156
Paprotny, I. 30–31, 156
Pareto, V. 136
Pascal, B. 37
Pasquali, V. 49, 151
Pearson, I. 93, 145
Pearson, M. J. 34, 154
Pellicani, L. 21, 67, 156
Petroni, A. M. 148, 157
Pirani, E. 41, 157
Plato 96, 147
Polchi, V. 50–52, 157
Polidori, A. 28, 29, 157
Popper, K. 147
Potter, S. 28, 95, 157
Poundstone, W. 135, 157
Proyas, A. 86
Putnam, H. 99, 157
Pythagoras 147

R
Rampini, F. 95, 157

Ranisch, R. 79, 157
Rao, S. 27
Ray, S. 134
Reagan, R. 132
Relke, D. M. A. 125, 157
Reutter, M. 15, 157
Ricardo, D. 14, 78, 155, 157
Rifkin, J. 56, 157
Riotta, G. 97, 157
Rogers, A. 110, 157
Rubart, J. 15, 157
Russo, L. 23, 67, 157
Russo, M. 41, 157

S
Sacconi, M. 50
Saint-Just, L. A. 105
Salvadori, N. 15, 158
Sanchez, M. 119, 158
Sandini, G. 27–29
Say, J. B. 14
Schor, J. B. 71, 158
Serreli, V. 145, 158
Sharkey, N. 16–17, 125–126, 135, 137, 158
Shin, D. 30, 158
Shoell, H. 140
Shue, H. 125, 153
Siciliano, B. 29–30, 39, 47, 158–159
Simmons, R. 34, 153
Singer, P. 16, 127–130, 132–134, 137, 139, 143–145, 158
Sismondi, J. C. 14
Sloggett, D. 110, 158
Sloterdijk, P. 97, 158
Smelser, N. 21
Smith, W. 86
Solow, R. 59
Sonnenberg, B. 15, 158
Sorgner, S. L. 21, 79, 157
Sparrow, R. 125, 159

Spencer, H. 11
Spiers, A. 34, 154
Sterling, B. 79
Stockhammer, E. 15, 159
Sullins, J. P. 125, 159
Sun, D. 31, 155
Sztompka, P. 21

T
Tabarrok, A. 57, 159
Tagliasco, V. 27
Takanisi, A. 79
Tan, K. C. 31
Tan, Y. K. 34
Tanie, K. 79
Thomas, D. S. 104
Thomas, W. I. 104
Tipler, F. 100, 102–103, 105
Travaglia, M. L. 41, 155
Tscheligi, M. 34, 160
Tucker, A. W. 135
Turing, A. 29, 127
Turse, N. 125, 159

U
Ura, T. 79

V
Verne, J. 128
Veruggio, G. 78–80, 82, 89, 159
Vltchek, A. 125, 152
Vroman, W. 15, 159

W
Wagenaar, D. A. 28, 157
Wang, L. 31, 159
Weber, M. 11–12, 159
Weinberger, S. 118, 159
Weiner, T. 142, 160
Weiss, A. 34, 160
Wells, H. G. 128
Werding, M. 15, 160

Whittle, R. 110, 160
Williams, M.-A. 34
Williams, R. 98
Winnefeld, J. A. 113, 116, 160
Witkowski, L. 21
Wittes, B. 125, 160
Woods, R. 89, 91, 93, 154

Y
Yeh, X. 30, 158
Young, J. J. 110, 152

Z
Zaloga, S. J. 111, 160
Złotowski, J. 34, 160
Znaniecki, F. 11, 160

BEYOND HUMANISM: TRANS- AND POSTHUMANISM
JENSEITS DES HUMANISMUS: TRANS- UND POSTHUMANISMUS

Edited by / Herausgegeben von Stefan Lorenz Sorgner

Vol./Bd. 1 Robert Ranisch / Stefan Lorenz Sorgner (eds.): Post- and Transhumanism. An Introduction. 2014.

Vol./Bd. 2 Stephen R. L. Clark: Philosophical Futures. 2011.

Vol./Bd. 3 Hava Tirosh-Samuelson / Kenneth L. Mossman (eds.): Building Better Humans? Refocusing the Debate on Transhumanism. 2012.

Vol./Bd. 4 Elizabeth Butterfield: Sartre and Posthumanist Humanism. 2012.

Vol./Bd. 5 Stefan Lorenz Sorgner / Branka-Rista Jovanovic (eds.). In cooperation with Nikola Grimm: Evolution and the Future. Anthropology, Ethics, Religion. 2013.

Vol./Bd. 6 Bogdana Koljević: Twenty-First Century Biopolitics. 2015.

Vol./Bd. 7 Riccardo Campa: Humans and Automata. A Social Study of Robotics. 2015.

www.peterlang.com

www.ingramcontent.com/pod-product-compliance
Ingram Content Group UK Ltd.
Pitfield, Milton Keynes, MK11 3LW, UK
UKHW041913140426
5217IPUK00002B/20